Solar Thermal Hot Water

RENEWABLE ENERGIES SERIES

skills2learn
www.skills2learn.com
Experts in e-learning & virtual reality simulation

CENGAGE
Learning

Australia • Brazil • Japan • Korea • Mexico • Singapore • Spain • United Kingdom • United States

Solar Thermal Hot Water, 1st Edition
Skills2Learn

Publishing Director: Linden Harris

Commissioning Editor: Lucy Mills

Development Editor: Lauren Darby

Senior Project Editor: Alison Burt

Senior Manufacturing Buyer: Eyvett Davis

Typesetter: MPS Limited

Cover design: HCT Creative

For product information and technology assistance,
contact **emea.info@cengage.com**.

For permission to use material from this text or product,
and for permission queries,
email **emea.permissions@cengage.com**.

British Library Cataloguing-in-Publication Data

A catalogue record for this book is available from the British Library.

ISBN: 978-1-4080-6468-9

Cengage Learning EMEA

Cheriton House, North Way, Andover, Hampshire, SP10 5BE
United Kingdom

Cengage Learning products are represented in Canada by Nelson Education Ltd.

For your lifelong learning solutions, visit **www.cengage.co.uk**

Purchase your next print book, e-book or e-chapter at **www.cengagebrain.com**

Printed in China by RR Donnelley
1 2 3 4 5 6 7 8 9 10 – 15 14 13

Contents

Foreword

The energy sector is a significant part of the UK economy and a major employer of people. It has a huge impact on the environment and plays a massive role in our everyday life, shaping both our work and domestic habits and processes. With environmental issues such as climate change and sustainable sourcing of energy now playing an important role in our society, there is a need to educate a significant pool of people about the future technologies, with renewable energies in all likelihood playing an increasingly significant part in our total energy requirements.

This environmental and renewable energy series of e-learning programmes and text workbooks has been developed to provide a structured blended learning approach that will enhance the learning experience, stimulate a deeper understanding of the renewable energy trades and give an awareness of sustainability issues. The content within these learning materials has been aligned as far as is currently possible to the units of the National Occupational Standards and can be used as a support tool whilst studying for any relevant vocational qualifications.

The uniqueness of this renewable energy series is that it aims to bridge the gap between classroom-based and practical-based learning. The workbooks provide classroom-based activities that can involve learners in discussions and research tasks as well as providing them with understanding and knowledge of the subject. The e-learning programmes take the subject further, with high quality images, animations and audio further enhancing the content and showing information in a different light. In addition, the e-practical side of the e-learning places the learner in a virtual environment where they can move around freely, interact with objects and use the knowledge and skills they have gained from the workbook and e-learning to complete a set of tasks whilst in the comfort of a safe working environment.

The workbooks and e-learning programmes are designed to help learners continuously improve their skills and provide a confidence and sound knowledge base before getting their hands dirty in the real world.

About e-Consortia

This series of renewable energy workbooks and e-learning programmes has been developed by the E-Renewable Consortium. The consortium is a group of colleges and organizations that are passionate about the renewable energy industry and are determined to enhance the learning experiences of people within the different trades or those that are new to it.

The consortium members have many years experience in the renewable energy and educational sectors and have created this blended learning approach of interactive e-learning programmes and text workbooks to achieve the aim of:

- Providing accessible training in different areas of renewable energy
- Bridging the gap between classroom-based and practical-based learning
- Providing a concentrated set of improvement learning modules
- Enabling learners to gain new skills and qualifications more effectively
- Improving functional skills and awareness of sustainability issues within the industry
- Promoting health and safety in the industry
- Encouraging training and continuous professional development.

For more information about this renewable energy consortium please visit: **http://skills2learn.cengage.co.uk/9-renewable-energy**

About e-learning

INTRODUCTION

This renewable energies series of workbooks and e-learning programmes uses a blended learning approach to train learners about renewable energy skills. Blended learning allows training to be delivered through different mediums such as books, e-learning (computer-based training), practical workshops and traditional classroom techniques. These training methods are designed to complement each other and work in tandem to achieve overall learning objectives and outcomes.

E-LEARNING

The Solar Thermal e-learning programme that is also available to sit alongside this workbook offers a different method of learning. With technology playing an increasingly important part of everyday life, e-learning uses visually rich 2D and 3D graphics/animation, audio, video, text and interactive quizzes, to allow you to engage with the content and learn at your own pace and in your own time.

E-ASSESSMENT

Part of the e-learning programme is an e-assessment 'End test'. This facility allows you to be self-tested using interactive multimedia by answering questions on the e-learning modules you will have covered in the programme. The e-assessment provides feedback on both correctly and incorrectly answered questions. If answers are incorrect the learner is advised to revisit the learning materials they need to study further.

E-PRACTICAL

Part of the e-learning programme is an e-practical interactive scenario. This facility allows you to be immersed in a virtual reality situation where the choices you make affect the outcome. Using 3D technology, you can move freely around the environment, interact with

objects, carry out tests, and make decisions and mistakes until you have mastered the subject. By practising in a virtual environment you will not only be able to see what you've learnt but also analyze your approach and thought process to the problem.

BENEFITS OF E-LEARNING

Diversity – E-learning can be used for almost anything. With the correct approach any subject can be brought to life to provide an interactive training experience.

Technology – Advancements in computer technology now allow a wide range of spectacular and engaging e-learning to be delivered to a wider population.

Captivate and Motivate – Hold the learner's attention for longer with the use of high quality graphics, animation, sound and interactivity.

Safe Environment – E-practical scenarios can create environments which simulate potentially harmful real-life situations or replicate a piece of dangerous equipment, therefore allowing the learner to train and gain experience and knowledge in a completely safe environment.

Instant Feedback – Learners can undertake training assessments which feedback results instantly. This can provide information on where they need to re-study or congratulate them on passing the assessment. Results and Certificates could also be printed for future records.

On-Demand – Can be accessed 24 hours a day, 7 days a week, 365 days of the year. You can access the content at any time and view it at your own pace.

Portable Solutions – Can be delivered via a CD, website or LMS. Learners no longer need to travel to all lectures, conferences, meetings or training days. This saves many man-hours in reduced travelling, cost of hotels and expenses amongst other things.

Reduction of Costs – Can be used to teach best practice processes on jobs which use large quantities of expensive materials. Learners can practice their techniques and boost their confidence to a high enough standard before being allowed near real materials.

SOLAR THERMAL E-LEARNING

The aim of the Solar Thermal e-learning programme is to enhance a learner's knowledge and understanding of solar domestic hot water installation and systems. The course content is aligned to units from the Environmental National Occupational Standards (NOS), so can be used for study towards certification.

The programme gives the learner an understanding of the different types of solar thermal panels, as well as looking at sustainability, health and safety and functional skills in an interactive and visually engaging manner. It also provides a 'real-life' scenario where the learner can apply the knowledge gained from the tutorials in a safe yet practical way.

By using and completing this programme, it is expected that learners will:

- Identify optimum conditions where SDHWS would be beneficial
- Identify factors which must be taken into account when SDHWS are installed and how to mitigate their effects
- Describe different types of solar collectors, solar primary circuits and solar secondary storage and distribution systems
- Calculate the size of collector and store required for a particular house
- Describe installation and methods used prior to commissioning a system
- Describe commissioning a system prior to handover to a client
- List maintenance procedures
- List fault finding location techniques and rectification of those faults

The e-learning programme is divided into the following learning modules:

- Getting Started
- Solar Thermal Overview and Considerations
- Health and Safety
- Solar Collector Types and British Standards
- Solar Thermal Storage Options
- Solar Thermal Primary Circuit Designs
- Commissioning
- Solar Primary Circuit Controls

- Installation Materials and Fittings
- Filling, Commissioning and Maintenance
- Interactive E-Practical Scenarios

THE RENEWABLE ENERGIES SERIES

As part of the renewable energies series the following e-learning programmes and workbooks are available. For more information please visit: **http://skills2learn.cengage.co.uk/9-renewable-energy**

- Introduction to Renewable Energies
- Heat Pumps
- Solar PV
- Building Heat Loss Calculator (programme only)
- Solar Radiation Calculator (programme only)

About the NOS

The National Occupational Standards (NOS) provide a framework of information that outline the skills, knowledge and understanding required to carry out work-based activities within a given vocation. Each standard is divided into units that cover specific activities of that occupation. Employers, employees, teachers and learners can use these standards as an information, support and reference resource that will enable them to understand the skills and criteria required for good practice in the workplace.

The standards are used as a basis to develop many vocational qualifications in the United Kingdom for a wide range of occupations. This workbook and associated e-learning programme is aligned to the Environmental National Occupational Standards (NOS) by being designed against the Qualification and Credit Framework (QCF) units, which are developed from the NOS. Such a process is a requirement of the minimum technical competency (MTC) document for solar installers. Therefore this book and its associated e-learning support full training towards a certificate as recognized by all bodies offering routes to the Microgeneration Certification Scheme (MCS) as evidence of suitable training. The information within relates to the following units:

- Know the health and safety risks and safe systems of work associated with solar thermal hot water system installation work
- Know the requirements of relevant regulations/standards relating to practical installation, testing and commissioning activities for solar thermal hot water system installation work
- Know the types and layouts of solar thermal hot water systems that incorporate a sealed collector circuit
- Know the purpose of components used within solar thermal hot water system installations
- Know the types and key operating principles of solar collectors
- Know the information requirements to enable system component selection and sizing
- Know the fundamental techniques used to select, size and position components for solar thermal hot water systems

- Know how the performance of solar hot water systems is measured
- Know how to and be able to undertake preparatory work required for solar thermal hot water system installation work
- Know the requirements for connecting solar thermal hot water system collector circuits to combination boiler domestic hot water circuits
- Know the requirements for installing solar collector arrays
- Know the requirements for installing solar thermal hot water system pipework
- Be able to install in accordance with manufacturers' guidance, regulatory requirements and industry recognized procedures for solar thermal hot water systems
- Know how to and be able to test and commission solar thermal hot water system installations
- Know the requirements to test and commission, and be able to test and commission solar thermal hot water system installations
- Know the requirements to handover solar thermal hot water systems and be able to handover solar thermal hot water systems

SUMMARY OF THE ABOVE

Or in simplified English, this book and the e-Learning training materials have been designed around the latest guidance from all the relevant bodies that will support a full solar thermal training course and assessment process.

They share the knowledge in a simple-to-digest format that is more enjoyable to use and so more likely to be successful in sharing the important information required to install, commission, handover and maintain solar thermal systems.

About the book

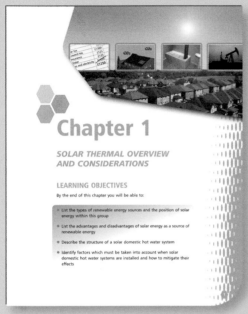

Learning Objectives at the start of each chapter explain the skills and knowledge you need to be proficient in and understand by the end of the chapter.

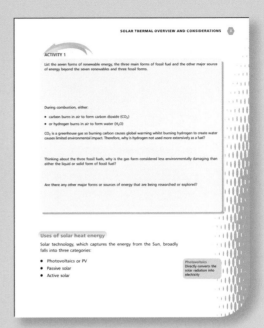

Activities are practical tasks that engage you in the subject and further your understanding.

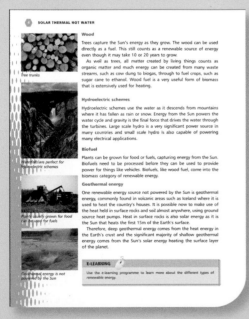

E-learning Icons link the workbook content to the e-learning programme.

Health and Safety Boxes draw attention to essential health and safety information.

Note on UK Standards draws your attention to relevant building regulations.

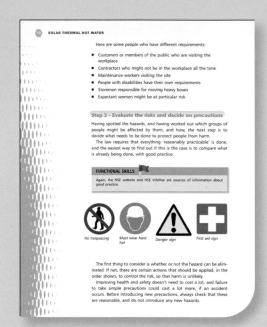

Functional Skills Icons highlight activities that develop and test your Maths, English and ICT key skills.

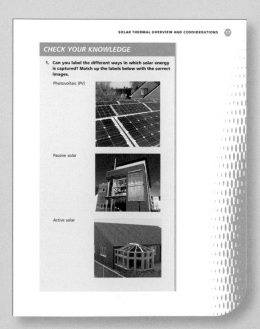

Check Your Knowledge at the end of each chapter to test your knowledge and understanding.

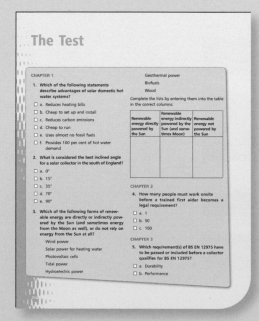

End Test in Chapter 9 checks your knowledge on all the information within the workbook.

Chapter 1

SOLAR THERMAL OVERVIEW AND CONSIDERATIONS

LEARNING OBJECTIVES

By the end of this chapter you will be able to:

- List the types of renewable energy sources and the position of solar energy within this group

- List the advantages and disadvantages of solar energy as a source of renewable energy

- Describe the structure of a solar domestic hot water system

- Identify factors which must be taken into account when solar domestic hot water systems are installed and how to mitigate their effects

Rooftop view of a suburban area

SOLAR DOMESTIC HOT WATER SYSTEMS

The solar domestic hot water system has four main purposes:

- To collect solar energy for heating water
- To reduce annual fuel bills
- To reduce our dependence on limited fossil fuels
- To cut carbon dioxide emissions

Collect solar energy for heating water

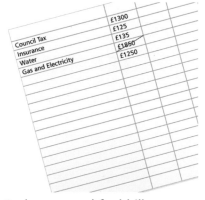

Reduce annual fuel bills

Reduce our dependence on fossil fuels

To reduce carbon dioxide emissions

Renewable energy

Reducing our dependence on fossil fuels can be achieved by turning to renewable energy. Solar energy is the most abundant and obvious, and besides tidal and geothermal, all the other renewables ultimately come from the Sun.

We can reducing dependency on fossil fuels by using renewable energy

> **Geothermal**
> Commonly found in volcanic areas such as Iceland where it is used to heat the country's houses

Wind power

Wind is a renewable source of energy caused by the Sun heating our atmosphere and creating our weather. The wind is the movement of air between a hot spot and a cooler location. Hot air rises and so creates the wind.

Wave power

Waves are a result of weather systems caused ultimately by the Sun's solar heating. Waves are created by the action of the wind on the surface of the ocean. This is why the biggest waves are found in the biggest oceans, as the wind has more opportunity to interact with the ocean's surface layer of water.

Tidal power is caused by the Moon and Sun exerting gravitational force on the oceans. This is why spring tides are the biggest because during a spring tide, the forces of the Sun and Moon act in unison.

Both wave and tidal power can be harnessed as renewable energy sources and there are a variety of innovative designs to do this task.

Wind turbines

Wave and tidal power

Tree trunks

Waterfalls are perfect for hydroelectric schemes

Plants usually grown for food can be used for fuels

Geothermal energy is not powered by the Sun

Wood

Trees capture the Sun's energy as they grow. The wood can be used directly as a fuel. This still counts as a renewable source of energy even though it may take 10 or 20 years to grow.

As well as trees, all matter created by living things counts as organic matter and much energy can be created from many waste streams, such as cow dung to biogas, through to fuel crops, such as sugar cane to ethanol. Wood fuel is a very useful form of biomass that is extensively used for heating.

Hydroelectric schemes

Hydroelectric schemes use the water as it descends from mountains where it has fallen as rain or snow. Energy from the Sun powers the water cycle and gravity is the final force that drives the water through the turbines. Large scale hydro is a very significant power source in many countries and small scale hydro is also capable of powering many electrical applications.

Biofuel

Plants can be grown for food or fuels, capturing energy from the Sun. Biofuels need to be processed before they can be used to provide power for things like vehicles. Biofuels, like wood fuel, come into the biomass category of renewable energy.

Geothermal energy

One renewable energy source not powered by the Sun is geothermal energy, commonly found in volcanic areas such as Iceland where it is used to heat the country's houses. It is possible now to make use of the heat held in surface rocks and soil almost anywhere, using ground source heat pumps. Heat in surface rocks is also solar energy as it is the Sun that heats the first 15m of the Earth's surface.

Therefore, deep geothermal energy comes from the heat energy in the Earth's crust and the significant majority of shallow geothermal energy comes from the Sun's solar energy heating the surface layer of the planet.

E-LEARNING

Use the e-learning programme to learn more about the different types of renewable energy.

ACTIVITY 1

List the seven forms of renewable energy, the three main forms of fossil fuel and the other major source of energy beyond the seven renewables and three fossil forms.

During combustion, either:

- carbon burns in air to form carbon dioxide (CO_2)

- or hydrogen burns in air to form water (H_2O)

CO_2 is a greenhouse gas so burning carbon causes global warming whilst burning hydrogen to create water causes limited environmental impact. Therefore, why is hydrogen not used more extensively as a fuel?

Thinking about the three fossil fuels, why is the gas form considered less environmentally damaging than either the liquid or solid form of fossil fuel?

Are there any other major forms or sources of energy that are being researched or explored?

Uses of solar heat energy

Solar technology, which captures the energy from the Sun, broadly falls into three categories:

- **Photovoltaics** or PV

- Passive solar

- Active solar

Photovoltaics
Directly converts the solar radiation into electricity

Solar PV panels

The energy from the sun falls into 3 categories; photovoltaic, passive and solar

Use of a lot of glass in a structure will heat the building

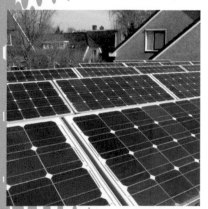
Solar collectors to gather maximum solar energy

Photovoltaics

Photovoltaics or PV directly convert the solar radiation into electricity.

Passive solar energy

In buildings with conservatories or south-facing entrance halls or similar structures with lots of glass, solar energy will heat the building. This is called passive solar radiation and with good design, can be used for space heating.

Active solar energy

Active **solar heating** involves a solar **collector** mounted in such a way as to gather the maximum amount of solar energy. This is mostly used in domestic hot water systems and for heating swimming pools. The domestic hot water system is the subject of this workbook.

ACTIVITY 2

In active and passive solar systems, the energy is normally collected by a fluid and this fluid is either pumped (active systems) or moves by gravity (passive systems). Please name the two most commonly used fluids (and please note that these fluids are all around us) in solar systems and also how the liquid form of the fluid is often treated to prevent it turning into a solid.

Active solar heating
Active solar heating involves a solar collector mounted in such a way as to gather the maximum amount of solar energy. This is mostly used in domestic hot water systems and for heating swimming pools. Active solar heating means that the system is pumped rather than gravity driven

Solar radiation in the UK

Different parts of the UK receive different levels of annual solar radiation. These variations are not only affected by the seasons and latitude, but also the position of hills, mountains and trees. And on top of these variables you also have differences created by the weather and how much sun is available. All these factors affect the amount of energy that can be generated per m² of a solar collector.

Here follows a discussion about Diffuse, Direct and seasonal variations in solar energy.

Collector The panel which collects solar energy and can be made from a variety of materials

Factors affecting solar energy levels:
The available energy decreases as you move north.
Local variations occur such as the presence of mountains or the coast, as well as the weather, which can cause up to 10 per cent annual variation in system output.

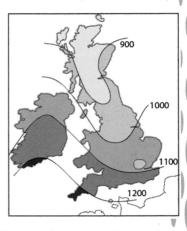

Figures shown on the map are average kWh/m² per year

Solar energy comes in two forms: the technical terms for these two forms are:

- Direct radiation
- Diffuse radiation

Direct radiation

Direct solar radiation is sunshine, which is strongest in the summer at midday because then the angle of the Sun is at its highest. There is therefore a daily variation in solar radiation as well as a seasonal variation between summer and winter. Direct radiation is the solar energy that travels in a straight line from the Sun to the planet's surface.

Direct solar radiation Direct solar radiation is sunshine, which is strongest in the summer at midday because then the angle of the Sun is at its highest

Zenith

21 June

21 September
21 March

21 December

W

N

S

E

04:00am

06:30am

08:33am

Direct solar radiation is sunshine

Diffuse radiation

Diffuse solar radiation gives us daylight and it can be quite strong. When the Sun's rays hit the Earth's atmosphere some radiation is scattered in all directions and is known as diffuse radiation. It is assumed to come from all over the sky equally, rather than just the direction of the Sun. As the UK has many overcast days more than 50 per cent of solar radiation reaching a horizontal surface at ground level is diffuse. This can be very different in countries such as those in Africa which can have over 90 per cent of their solar energy as direct and so less than 10 per cent as diffuse.

Diffuse solar radiation is daylight, typically a cloudy day, or the daylight coming in a north-facing window. This type of solar radiation can still be used to power domestic hot water systems.

Diffuse solar radiation gives us daylight and can be quite strong

Seasonal radiation

Seasonal variation is easy to see – and feel – in the UK. The proportion of diffuse and direct radiation for each year can vary, as well as the total amount of solar radiation. This graph shows recordings for a poor summer and hot September, however it still shows over five times as much total radiation falling during the summer months compared to winter. Most solar radiation is received during the summer in the UK.

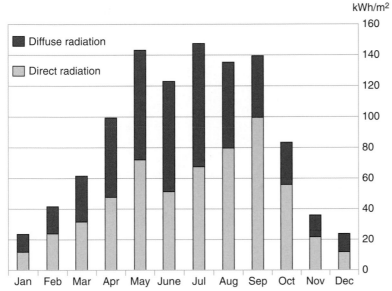

Graph showing the diffused and direct radiation throughout the year

E-LEARNING

Use the e-learning programme to learn more about the different types of solar energy.

ACTIVITY 3

What 'disorder' is associated with the lack of sunshine in winter and name two simple solutions that are available for this disorder. What vitamin comes from the Sun?

Solar orientation

Inclination The tilting of something away from a line or surface, or the degree to which it is tilted

The collectors' orientation and angle of **inclination** towards the Sun can make a significant difference to their efficiency. The graph shows how this can be worked out.

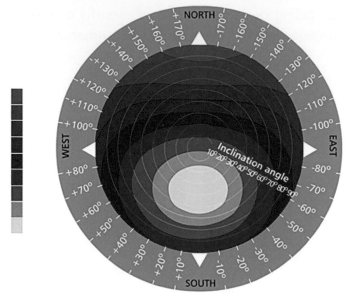

Solar orientation

Collector orientation

In the UK and across northern Europe collectors should ideally face south, which receives most solar radiation. However, orientations between 30° east and 40° west of south are acceptable and will not lose more than ten per cent in efficiency compared with the ideal situation. Alternatively the surface area of the collector could be increased to compensate for lower efficiency if outside these geographical positions.

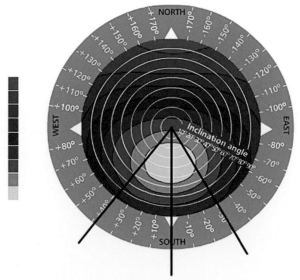

The graph shows what orientation and angle of inclination gains most efficiency

Collector angle of inclination

Collectors also need to be inclined towards the Sun, and an angle of 35° to the horizontal is best for the UK and across northern Europe. The graph shows there is some leeway in angles depending on the pitch of the roof. As can be viewed on the graph, for a collector facing directly south, a collector at an angle of 5° or 65° will not lose more than ten per cent, as compared to the optimum 35°.

If the roof is facing east or west and so is not at the optimum angle, a solar thermal system can still be fitted. In this situation, the collector should be 'oversized' to compensate for the loss in solar energy. This is discussed again in Chapter 5 of this book.

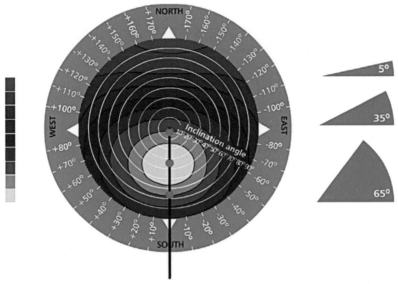

Inclination towards the Sun at an angle of 35 degrees is best

E-LEARNING

Use the e-learning to learn more about solar orientation.

ACTIVITY 4

Thinking about typical houses, would they typically have a pitched roof facing somewhere between south-east to south-west or east to west, and would the pitched roof typically have an optimal inclination?

SOLAR DOMESTIC HOT WATER SYSTEM BASICS

What makes up a solar domestic hot water system?

At its simplest, a solar domestic hot water system consists of a solar collector, a hot water storage vessel, heat transfer pipes, a means of circulation and a means of managing various factors that can affect the system. Heated fluid from the collector circulates to the cold water section of the storage vessel and heats the water. The pump keeps the circulation going. Because solar radiation varies it cannot be relied upon to heat the water at all times, therefore other methods of heating water are also needed.

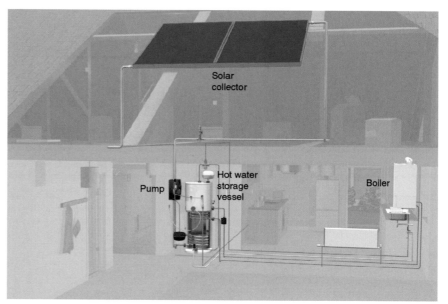

Components of a solar domestic hot water system

ACTIVITY 5

What other methods as well as solar power could be used to back up or support the solar energy so that hot water can be guaranteed for all the year? Solar energy tends to be strong in summer and mild in winter. Please explain when the solar system will need the most support or back up.

Solar system hazards to cover

Whilst a solar system collects solar energy, there are many features and design parameters that need to be addressed to design and install an effective solar system.

We can now look at these issues that need addressing in a bit more detail. Some issues can be solved, others need to be minimized so the efficiency and safe operation of the installed system is maximized. Study each issue before moving on.

Issues that need addressing, mitigating or solving when using a solar hot water system

When using a solar hot water systems there are issues to consider

Freezing

Frost protection is essential for the collector and surrounding pipes. Insulation alone is not sufficient to protect against frost. Some systems use antifreeze to protect against freezing and some systems can be drained to protect the pipes in the collector and nearby.

Collectors and surrounding pipes must have frost protection

High temperature and steam

Stagnation mode
When there is no fluid movement, sometimes in summer, the controller switches the solar pump off

When there is no fluid movement, i.e., the collector is in **stagnation mode**, sometimes in summer, the controller switches the solar pump off. In this stagnation mode, the solar collectors are capable of reaching between 150°C and 300°C and converting the transfer liquid to steam. The components must be able to withstand vaporization and condensation cycles, but also capable of operating without failure or distortion. Care must be taken with any work near the system to avoid scalding the installer, particularly when testing a new installation.

> **Stagnation** is the term used to describe what happens when the solar store has been fully charged during a long period of sunlight and the solar controller needs to switch the system off. When this happens, there is nowhere for the solar energy to go and so any liquid in the solar collector turns to steam. The solar system needs to manage this steam without the need for recharging the system when it cools down. This is not easy as water expands 1600 times when it changes to steam.

When in stagnation mode the controller switches the solar pump off

Limescale

Limescale can significantly reduce heat transfer and so you need to find out the supply water hardness. Test kits or local authority reports are good methods. The answer can include maintaining a temperature below 65°C, preferably at 60°C, or cleaning out the limescale physically. Water softeners will also reduce limescale. However, the energy saved can be offset by the energy consumed by the water softener.

Limescale, as well as reducing system efficiency, also acts as a good harbour for bacteria and so **Legionella** and other health **risks** increase in areas of hard water. However, it is said to make the water taste better and drinking hard water is also said to be good for human health as the salts in the water are good for the body.

Legionella A disease that can form in pipes and that likes temperatures around 20°C to 50°C, particularly around 38°C to 40°C, with other factors such as nutrients and being in slow moving or stagnant fluids. Copper pipes and chlorine are toxic to Legionella. This disease can kill

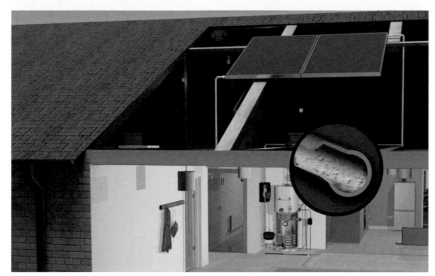

Limescale can significantly reduce heat transfer

Risk Refers to how likely it is that a potential hazard will actually damage your health

Bacterial growth

Conditions within the system should not offer favourable conditions for bacterial growth – especially Legionella. Legionella likes temperatures around 20°C to 50°C, particularly around 38°C to 40°C, with other factors such as nutrients, i.e. rubber components, limescale, biofilms and foreign objects such as insects, and being in slow moving or stagnant fluids.

Unfavourable conditions include higher temperatures, i.e. temperatures above 50°C (sterilize bacteria in hours) or 60°C (sterilize bacteria in minutes) or 70°C (sterilize bacteria in seconds) water flow, no 'dead legs' of pipework, regular cleaning and copper pipes or chlorine; both are toxic to Legionella.

Scalding and Bacteria

All hot water systems should have a scalding and bacteria **risk assessment** covered before the installation goes ahead. Scalding is a major issue which can cause death, especially in babies and older people. Therefore, blending valves and other measures should be installed as appropriate. Legionella and bacteria pasteurization and protection is also a significant risk and a regular pasteurization cycle should be implemented. The bacteria risks should be assessed and as appropriate reduced upstream, inside and downstream of the hot water cylinder.

Risk assessment Identifying hazards in the workplace then deciding who might be harmed and how

Bacterial growth e.g. Legionella

ACTIVITY 6

In this last section, we discussed freezing, stagnation, limescale, bacteria and scalding as issues to address and manage in a solar water heating system. Thinking about these issues, how are they addressed in a condensing gas boiler-based heating system, both in the central heating primary pipework and the secondary hot water distribution system?

CHECK YOUR KNOWLEDGE

1. **Can you label the different ways in which solar energy is captured? Match up the labels below with the correct images.**

Photovoltaic (PV)

Passive solar

Active solar

2. **When a solar hot water system is designed for a building what issues do you think need to be addressed and solved before work can go ahead?**

☐ a. Solar fluid expansion

☐ b. Drought

☐ c. High solar collector temperatures

☐ d. Scalding

☐ e. Stagnation

☐ f. Bacteria growth (Legionella)

☐ g. Limescale

☐ h. Crystallization

☐ i. Short trees nearby

☐ j. Freezing

3. **Are the statements for each hazard True or False?**

Hazard	Solution	Answer
Steam & scalding	The components must be able to withstand vapourization and condensation cycles, but also capable of operating without failure or distortion	
Frost & freezing	Provide low level heating to prevent ice forming in pipes	
Bacterial growth	Make sure the temperature of the water is raised to 45°C at least once a day	
Limescale	**SDHWS** can only be installed in soft water areas as limescale will not form in the pipework	

SDHWS Solar Domestic Hot Water System

4. **When will there be sufficient solar energy to provide some or all of the heating for a domestic hot water system in the UK?**

☐ a. A sunny day in summer

☐ b. At night in autumn

☐ c. A fairly cloudy day in spring

☐ d. A bright, cold day in winter

5. **On the map shown, split the UK into 5 areas based on the level of annual solar radiation they receive. Also rate each area in terms of lowest radiation to highest radiation on a scale of 1 to 5.**

Chapter 2

HEALTH AND SAFETY

LEARNING OBJECTIVES

By the end of this chapter you will be able to:

- List the key items of Personal Protective Equipment

- Identify common safety and hazard signs

- Describe the effects of hazardous substances

- Describe safe working at heights

- Identify the key elements of a first aid kit

- Know how to carry out a risk assessment

- Match fire extinguishers to different types of fire

- Select CSCS safety cards for different types of people

- Describe all aspects of electrical safety

Worker wearing correct PPE while at work

HEALTH AND SAFETY AT WORK ACT

HASAWA 1974

> **Health and Safety at Work Act (1974) (HASAWA)** All employers are covered by the HASAWA, which places specific duties on both employers and employees to ensure that workplaces are safe. Non-compliance can result in fines

The **Health and Safety at Work Act, 1974** provides the legal framework to:

- promote
- stimulate
- encourage

high standards of health and safety in the workplace.

It protects employees and the public and puts the onus on everyone to be responsible for their health and safety and the health and safety of others who could be affected.

Before going any further though, it is essential that you have an understanding of the Health and Safety at Work Act, 1974.

The Act provides the legal framework to promote, stimulate and encourage high standards of health and safety in the workplace. It protects employees and the public from work activities.

Everyone has a duty to comply with the Act, including employers, employees, trainees, the self-employed, manufacturers and suppliers.

Workers must comply with the HASAWA

If negligence can be proved, both employers and employees can face a £5000 fine from a Magistrate's Court and unlimited fines and imprisonment from a Crown Court.

Employer responsibilities

The Act places a general duty on employers to ensure 'so far as is reasonably practicable the health, safety and welfare at work of all their employees'.

Employers must provide and maintain safety equipment and safe systems of work. This includes, among other things, ensuring materials are properly stored, handled, used and transported, providing information, training, instruction and supervision, and ensuring staff are aware of manufacturers' and suppliers' instructions for equipment. Employers must also look after the health and safety of others, for example the public, and talk to safety representatives.

Employers are forbidden to charge employees for any measures which are required for health and safety, for example, personal protective equipment.

Employers must ensure the health, safety and welfare of employees

Employee responsibilities

Employees must comply with the Act too and look after their own health and safety as well as the health and safety of others. They must co-operate with their employers and not interfere with anything provided in the interest of health and safety.

Employees must comply with the act - look after their health and safety

PPE

PPE Personal Protective Equipment

'PPE' stands for 'Personal Protective Equipment' and covers a range of different items of clothing or equipment – such as gloves or safety helmets – that you may have to use on a job to avoid harm or injury.

Employers have a legal duty to identify any risks involved with a particular job and thus what items of PPE may be needed – but you still need to know the basics so you can make sure you get what you need.

As ideally risks to health and safety should be eliminated from the workplace before they occur, PPE is a last line of defence. However, if risks remain, then PPE must be provided to you free of charge.

Worker wearing the correct PPE

Common PPE items

You need to be familiar with several key items of PPE. You may not need to use them on every job, but you still need to know when they are required and how they should be used.

Commonly used PPE

Safety footwear

Safety footwear

For many jobs you will need to wear steel-toed boots with intersoles – these are thin pads in the boots or shoes that absorb surface shock – to protect your feet from injury.

If working in wet conditions, rubber boots should be worn – which must also have intersoles and steel toecaps.

Overalls

Overalls

Boiler suits are ideal for many jobs as they provide cover for your entire body. However, you must never wear overalls made from terylene, nylon or similar materials as these catch fire easily.

Ear muffs

Ear muffs/ear plugs

If you are exposed to high noise levels, you must protect your hearing with ear muffs or ear plugs.

Ear muffs must always be properly fitted so that the ear is completely covered, otherwise you will not be fully protected.

Ear plugs fit inside the ear, and are often disposable. They offer less protection than ear muffs.

Respirators

Many different types of **respirator** are available, but filter masks are the most common. These are rubber face masks that fit over the mouth and nose, containing a filter canister through which the wearer breathes.

Respirators

Filter masks stop dust but are useless against gases or vapours – so if you are working around these, you must use a canister respirator. Filter canisters must be changed regularly.

Respirator Filter masks stop dust but are useless against gases or vapours, for which you must use a canister respirator

Safety gloves

Safety gloves protect your hands from injury. Take care to choose the right kind, as different types exist, for example for working with heat or chemicals. Check glove application data, if available.

Do not wear gloves when using machinery – e.g. drills – as this is dangerous.

Safety gloves

Safety goggles

You must wear safety goggles when required, for example when welding, when working in dusty conditions, or when flying chippings will be produced.

Safety goggles

Always check that you have the right type of goggles, as different lenses offer different levels of protection.

If goggles have dirty lenses, clean them before use: never obscure your vision.

Safety helmets

Hard hats should be worn on a building site whenever there is a risk of falling or flying objects. You must always check that your helmet fits properly. Always wear your hat with the peak facing forward, as the peak lip is designed to protect your eyes. When working indoors, a 'bump hat' can be worn instead.

Safety helmet

Hard hats sometimes come with an expiry date. It is not advisable to use such hard hats after the expiry date as it may have deteriorated and not provide the necessary protection.

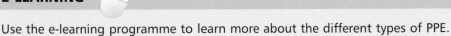

E-LEARNING

Use the e-learning programme to learn more about the different types of PPE.

SAFETY SIGNS

Risks and hazards

Work sites make extensive use of safety signs to warn of a number of different risks and hazards. You will learn what the different signs mean, but before you do it is important to understand what is meant by a 'risk' and a 'hazard' – as they are not the same thing.

Must wear hard hat

Safety sign

Hazards

A hazard exists if a substance can potentially damage your health. The negative effects of a hazard may be relatively minor – such as making your eyes water – or they may be much more serious, such as suffocation.

In some cases, they can be fatal. Hazards may be difficult to detect, and need not affect you immediately, such as the cancers eventually caused by asbestos over the long term.

Risks

Risk refers to how likely it is that a potentially hazardous substance will actually damage your health. The better controlled a hazardous substance is, and the more rigorous people are about using PPE properly, the lower the risks will be.

Identifying safety signs

To be able to work safely you must understand what the basic safety signs mean.

'**Prohibition**' **signs**, which are circular with red crosses through them, tell you not to do something.

Prohibition signs
These are circular with red crosses through them; they tell you not to do something

No pedestrians   *No smoking*   *No lit matches*   *Do not extinguish with water*

'**Mandatory**' **signs**, which are blue circles with white symbols, tell you what you must do.

Mandatory signs
These are blue circles with white symbols; they tell you what you must do

Must wear safety goggles   *Must wear safety gloves*   *Must wear ear muffs*   *Must wear hard hat*

Most signs usually have accompanying textual information to further explain the sign.

'Warning' signs have yellow triangles with black symbols, these give notice of a particular hazard or danger.

'Information' signs are used to communicate safety information. These are green squares with white symbols.

Emergency phone

Emergency exit direction

First aid

Warning sign in use on a building site

E-LEARNING

Use the e-learning programme to learn more about the signs and what each one means.

CONTROL OF SUBSTANCES HAZARDOUS TO HEALTH (COSHH)

Classifying hazardous substances

Hazardous substances can be broken down into four main categories:

- Toxic/very toxic substances
- Corrosive substances
- Harmful substances
- Irritants

Being able to identify these in a practical setting will help to alert you to possible sources of danger so that you can take appropriate protective action.

Dangerous chemicals

Dangerous acid

Danger

Irritant

Toxic/very toxic substances

Toxic/very toxic substances such as bleach can cause death or serious damage when inhaled, swallowed or absorbed by the skin.

Corrosive substances

Corrosive substances such as sulphuric acid may destroy parts of your body if they come into direct contact with them.

Harmful substances

Harmful substances such as lead can cause death or serious damage when inhaled, swallowed or absorbed by the skin.

Irritants

If irritants – such as soft solder flux – come into contact with the skin, eyes, nose or mouth, they can cause inflammation or swelling. Such effects may be felt immediately, or they may come after extended or repeated contact. Approach roofs on site with caution.

E-LEARNING

Use the e-learning programme to learn more about hazardous substances.

UK AND INTERNATIONAL STANDARDS

Under European law, substances are defined as being officially hazardous to health if they are listed as being 'dangerous to supply'.

They must also be defined as being very toxic, toxic, corrosive, harmful or irritants. One practical result of this is that when sold commercially, the packaging for such substances must be clearly marked and labelled.

Warning signs These warning signs are yellow triangles with black symbols and give notice of a particular hazard or danger

Hazard signs

Hazardous substances are identified by four main **warning signs**.

Explosion risk

Flammable materials

Toxic

Corrosive

Substances hazardous to health

Corrosive material

Substances marked as being 'corrosive' could cause permanent damage if they come into direct contact with any part of your body. For example, sulphuric acid will burn your skin away and cause breathing problems.

Flammable material

Flammable materials must be kept away from naked flames, and you should not smoke when near them. Common flammable materials include your own clothes, hair that is worn long, some modern hair products, and any oily rags that may have been left lying around.

Explosive material

Explosive materials must be handled and stored with particular care, as they potentially present an extreme hazard.

Toxic material

Some toxic materials – such as gas in a confined area – can harm you even if you do not come into direct contact with them, so ensure you handle them with care. If necessary, seek advice as to whether PPE is required when they are in use.

The effects of hazardous substances

In order to be able to adequately follow safety procedures and use PPE correctly, it is important to understand exactly how harmful substances can affect your body.

Skins and eyes

Some substances will cause damage – such as burns – if they come into direct contact with the skin. Contact may also lead to less serious problems, such as increased skin sensitivity. Harmful substances

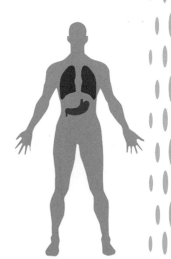
Hazardous substances can damage: skin, eyes, lungs and stomach

such as solvents can also enter the bloodstream by being absorbed through the skin, or through cuts and bruises.

Lungs

Harmful substances can cause damage when inhaled in one of two ways. Either they can cause direct damage to the lungs themselves in the way that asbestos does, or they can enter the bloodstream and so affect other organs.

Stomach

Harmful substances such as lead can reach the stomach in numerous ways if basic hygiene is not observed, or if gloves are not worn. Eating, drinking, smoking and biting your nails can all be responsible for this.

E-LEARNING

Use the e-learning programme to learn more about the effects of hazardous substances.

WORKING AT HEIGHTS

Using ladders safely

Take care when using ladders

People injure themselves using ladders for four main reasons. They overreach or slip, or the ladder itself breaks or falls over. These risks can be minimized if you stay aware of how to use ladders properly.

Leaning ladders

Before you use a leaning ladder, check that it has clean rungs and undamaged stiles (these are the side pieces the rungs are attached to). Position it at an angle of 75° from the vertical (e.g. 1m out from the base for every 4m of height).

Ensure that it can't move about at the top, that it has a strong upper resting point, and that its rungs are horizontal. When in use, never stand on the top three rungs, and maintain three points of contact at all times.

Position a leaning ladder at a 75 degree angle for the vertical

Stepladders

Before you get started, check that the stepladder is in good condition, with clean treads and secure locks. It should be fully open, and locked firmly in place. Do not work sideways on.

Ensure that the stepladder is in a good condition

Before you start – general principles

Whatever type of ladder you are using, before you start you should check that it is in good general condition. Are its feet firmly attached?

Have you properly secured its fastenings? You must also ensure that it is in a good position. It should not be able to move at the bottom, and should be placed on a firm, level surface that is clear and dry.

Finally, you must always be fully fit to work at heights, and never paint wooden ladders as this may conceal damage or cracks.

Ensure that the feet of the ladder you are using are firmly attached

Using ladders – general principles

You should never work on a ladder for longer than 30 minutes, and must not exceed stated weight limits – so only carry light tools and materials.

When working, keep your body centred on the ladder, avoiding overreaching. You must also keep both feet on the same rung or step. When you climb, always keep a firm grip on the ladder.

Non-slip footwear should always be worn.

When climbing, always keep a firm grip on the ladder

Safety boots must be worn at all times

Safety gloves must be worn

Avoid working for longer than 30 minutes on the ladder

Using scaffolding safely

Setting up, using and checking scaffolding must all be done with great care, otherwise falls and collapses can cause serious injury, or even death. Also, due consideration must be paid to public safety.

Great care must be taken when using scaffolding to avoid serious injury

Erecting scaffolding

Scaffolding must always be erected by competent people. It should be put up on firm, level ground away from any power lines. Uprights should be placed on base-plates.

Gaps are dangerous, so it should be no more than 300mm away from the building or structure it is attached to – and such attachments must be secure.

Working platforms must always be wide enough to allow people to use equipment and pass each other in safety – they should be no less than 600mm wide, and free of openings.

Using scaffolding

To avoid falls, strong barriers such as guardrails and toe boards must be in place. Heavy or bulky loads should never be carried up or down ladders, which should be strong, secure and in good condition. The top of the ladder should project 1m above the level of the work platform.

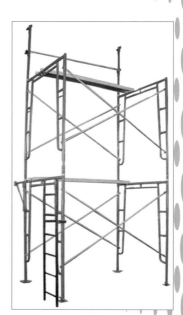

Avoid using domestic ladders, as they may collapse

The use of domestic ladders should be avoided, as these may collapse. Under no circumstances should components ever be removed from the scaffolding after it has been set up.

Risk to the public

Scaffolding should never normally be set up over busy public areas. If a risk exists, limit work to quiet times. Prevent waste materials from falling on people by keeping work platforms clear of debris – if they are allowed to build up, they may become stacked above the level of the toe boards.

Always avoid using scaffolding over busy public areas

Checking

Once it has been set up, scaffolding should be checked on a weekly basis, as well as after any alterations, damage, or extreme weather. Such inspections must be done by a competent person. Any problems should be fixed immediately, or the scaffolding must be taken out of use.

Scaffolding should be checked on a weekly basis

Tower scaffolds

Before use, you must ensure the tower is braced, has outriggers, and that any wheels are locked. Tower scaffolds must be kept away from overhead **cables**, and ladders should never be leant against them. You should never move a tower when people or heavy items of equipment are on board. Finally, as with ladders, you should avoid overreaching when you are standing on one.

Ensure the tower is braced, has outriggers and wheels are locked

E-LEARNING

Use the e-learning programme to learn more about using scaffolding safely.

> **Cable** A conductor used to carry current around an installation. Cables are identified by the colour of the installation

BASIC FIRST AID

However well prepared people are, accidents will happen. Employers have a legal duty to ensure that first aid kits are in place, containing sufficient supplies for all workers onsite.

For smaller work sites, clearly-marked first aid boxes must be placed under the control of a named individual. However, for larger sites of over fifty people, there must also be at least one person with first aid training.

First aid kit

First aid sign

MANUAL HANDLING

Manual handling includes lifting, carrying, lowering, pushing and pulling. We will consider lifting first, which can be broken down into five steps. The first is stop and think, when you plan the lift and assess the risks.

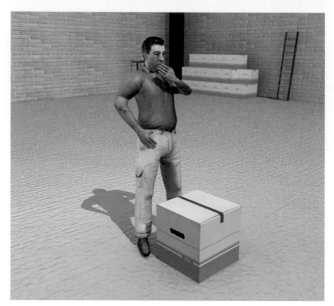

First step, stop and think

Lifting

1. Stop and think – plan the lift and assess the risks

2. Take up a good posture to start the lift and keep your back naturally straight

3. Take a firm grip on the load, no twisting or overreaching

4. Start to lift using power of the legs to make a smooth and controlled action. If the load is too heavy do not continue

5. Keep the load under control, heaviest side closest to the body and beware of liquids and uneven loads

Step 1

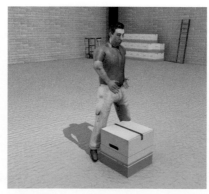

Step 2

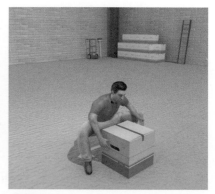

Step 3

Step 4

Step 5

Carrying

Carrying the load is the main purpose of manual handling. When carrying the load move your feet steadily and slowly, look ahead (not at the load) and avoid twisting the body.

- Keep the load close to maintain control
- Be vigilant and aware of your surroundings – especially be alert to possible danger, in addition to original assessment of risks. Be ready for the unexpected
- If feeling tired or strained – stop. Don't overdo it

Keep load close to you to maintain control

Pushing

For loads that are too large to be carried by one or two people, then an alternative must be found. Options available include trolleys, wheelbarrows and scissor lifts.

- Always check the trolley or barrow is in good working order
- Assess the whole route for hazards, especially slopes up or down and uneven surfaces
- Use safe lifting techniques to load the trolley
- Take a firm grip on the handles, take the weight of the load, lean forward slightly keeping a natural straight back
- Use the power of the legs to push in an even and controlled way, about slow walking pace
- Once moving, keep a steady pace and keep feet away from the load
- When you arrive at the destination slow down smoothly (no jerking movements) and unload in the same way it was loaded
- Always return the manual handling aid to its place to avoid turning it into a hazard, or preventing someone finding it when needed. If the aid cannot be found they may be tempted to do unsafe manual handling

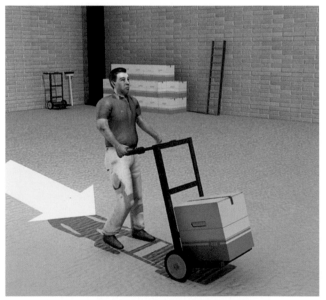

When the load is too heavy use trolleys, wheelbarrows or scissor lifts

Pulling

When you have a load on a trolley, or other manual handling aid, you can push or pull it. However, pulling is more dangerous than pushing as the load can 'run over' you. If you lose control when pushing the load will move away from you; if, however, you were pulling, the load will run towards you! Pulling also puts more strain on the back and it is less easy to control the load.

- Always check the trolley or barrow is in good working order
- Assess the whole route for hazards, especially slopes up or down and uneven surfaces
- Use safe lifting techniques to load the trolley
- Take a firm grip on the handles, take the weight of the load, lean backward slightly keeping a natural straight back
- Maintain a steady pace – about slow walking pace and keep feet away from the load
- When you arrive at the destination slow down smoothly (no jerking movements) and unload in the same way it was loaded
- If uncomfortable at any time stop and reassess the situation. You may need to lighten the load or get help

If you have a heavy load you can either push or pull it

Pointers to good manual handling practice

Carrying the load is the main purpose of manual handling. When carrying a load move your feet steadily and slowly, look ahead and not at the load, and avoid twisting the body. Keep the load close to maintain control – especially if the load is a difficult one such as a liquid or odd shape. Be vigilant and aware of your surroundings and especially be alert to possible danger. Your original assessment of risks was taken at a particular time – things may have changed so be ready for the unexpected! Remember, if you are feeling tired or strained – stop. Don't overdo it.

The main purpose of manual handling is carrying the load

ASSESSING HEALTH AND SAFETY RISKS IN THE WORKPLACE

Many accidents in the workplace should simply not happen. This is why it's very important to assess risks in the workplace, in order to protect not only the people working there, but also members of the public.

There is a legal requirement to carry out regular risk assessments, and, although the law does not expect all risks to be eliminated, people must be protected as far as it is reasonably practicable to do so. It is very important to regularly assess equipment in the workplace. The equipment shown below and on the following page are some of the key items that must be regularly risk assessed.

It's very important to assess the risks in the workplace

Tool box

Electrical equipment

Leaning ladder

Tower scaffolds

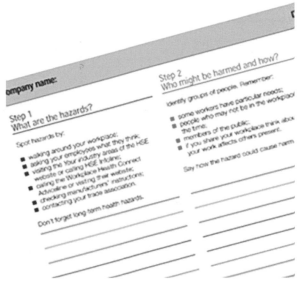

Health and safety forms must be completed

What is risk assessment?

Risk assessment means identifying hazards in the workplace then deciding who might be harmed and how.

A hazard can be anything that might cause harm, for example, electricity, chemicals, working from ladders, or an open drawer.

There are five steps in the risk assessment process. Once this process has been completed suitable precautions can be put in place to reduce the risk of harm, or make the harm less serious.

The five steps of risk assessment are:

- Identifying the hazards
- Identifying who could be affected
- Evaluating the risks
- Recording findings and implementing them
- Reviewing risk assessment

A hazard could be anything that may cause harm e.g. chemicals

Step 1 – Identify the hazards

The first step is to identify what the hazards are. When you work in a place every day, it's easy to overlook potential hazards, so there are ways to make sure you identify those that matter.

Walking around to look for things that might reasonably be expected to cause harm is a good starting point. Also, ask employees or their representatives, as they might be aware of things that are not immediately obvious to you.

Publications and practical guidance are available from a variety of sources, including the Health and Safety Executive, or HSE, and relevant trade associations, as well as manufacturers' instructions. These all provide information on where hazards can occur, their harmful effects, and how to control them.

Accident and ill-health records are another source of information which can often help to identify the less obvious hazards. Don't forget the long-term health hazards, for example, those that can occur following prolonged exposure to high levels of noise, or harmful substances.

Working in the same place every day can make it easy to overlook certain hazards, for example, leaving equipment unattended

Incorrect manual handling

Electrical safety hazard

Items left lying about could cause accidents

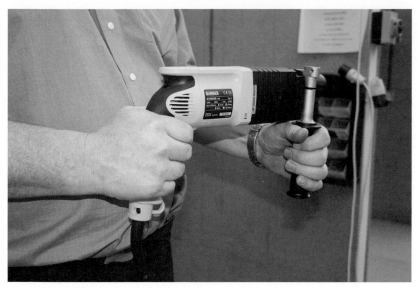

Long-term exposure to excessive noise

Here are some hazards:

- Trailing cables might cause people to trip
- Spills might cause people to slip and may also be toxic
- Faulty or damaged electrical fittings might cause electric shock and possibly a fire
- Items left lying about might cause people to trip or block access
- Lifting heavy loads might cause back injuries
- Long-term exposure to excessive noise might cause loss of hearing

Step 2 – Who might be harmed and how

The second step in risk assessment starts with deciding who might be harmed by each of the identified hazards.

To do this, it's best to identify groups of people, rather than listing people by name, as this will help later on when it comes to identifying the best ways to manage the risk for each group.

It's also necessary to identify how the different groups of people might be harmed; for example, what type of injury or ill-health might occur.

A customer or member of the public who is visiting the work place must make themselves known to the project manager

Maintenance workers must make visitors aware of hazards like slippery floors

Storemen responsible for moving heavy boxes

Contractors who may not be in the workplace at all times

People with disabilities have their own requirements

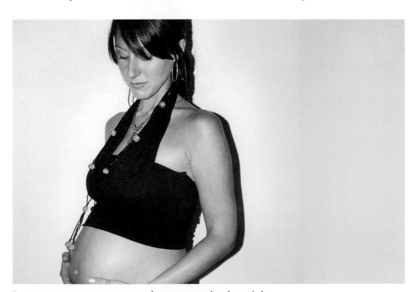

Pregnant women may be at particular risk

Here are some people who have different requirements:

- Customers or members of the public who are visiting the workplace
- Contractors who might not be in the workplace all the time
- Maintenance workers visiting the site
- People with disabilities have their own requirements
- Storemen responsible for moving heavy boxes
- Expectant women might be at particular risk

Step 3 – Evaluate the risks and decide on precautions

Having spotted the hazards, and having worked out which groups of people might be affected by them, and how, the next step is to decide what needs to be done to protect people from harm.

The law requires that everything 'reasonably practicable' is done, and the easiest way to find out if this is the case is to compare what is already being done, with good practice.

FUNCTIONAL SKILLS

Again, the HSE website and HSE Infoline are sources of information about good practice.

No trespassing   *Must wear hard hat*   *Danger sign*   *First aid sign*

The first thing to consider is whether or not the hazard can be eliminated. If not, there are certain actions that should be applied, in the order shown, to control the risk, so that harm is unlikely.

Improving health and safety doesn't need to cost a lot, and failure to take simple precautions could cost a lot more, if an accident occurs. Before introducing new precautions, always check that these are reasonable, and do not introduce any new hazards.

Record the risks found

Contracts

Use HSE website for information

**HSE Infoline:
0845 345 0055**

Call the HSE infoline if in doubt

Here are some different actions that could be applied:

- Switch to using a less hazardous method
- Switch to using a less hazardous chemical
- Prevent access by guarding the hazard
- Provide lifting equipment
- Provide clothing, footwear, goggles, etc.
- Provide first aid and washing facilities for removal of contamination

Step 4 – Record findings and implement them

After you've spotted the hazards, worked out which groups of people might be affected by them, and how, and decided what needs to be done to protect people from harm, it's important to keep a record of what's been done.

In fact, for businesses with five or more employees, the results of the risk assessment must be written down, and actions recorded as they are implemented.

Smaller businesses will also find it useful to have a written record of their risk assessment, as it can be reviewed at a later date.

The main thing is to keep the written results of the risk assessment as simple as possible, and to share the document with employees.

Company name:	
Step 1 What are the hazards?	**Step 2** Who might be harmed and how?

Spot hazards by:
 	
■ walking around your workplace;
■ asking your employees what they think;
■ visiting the Your industry areas of the HSE
   website or calling HSE Infoline;
■ calling the Workplace Health Connect
   Adviceline or visiting their website;
■ checking manufacturers' instructions;
■ contacting your trade association.
 	
Don't forget long-term health hazards.
	Identify groups of people. Remember:
	
	■ some workers have particular needs;
	■ people who may not be in the workplace all
	   the time;
	■ members of the public;
	■ if you share your workplace think about how
	   your work affects others present.
	
	Say how the hazard could cause harm.

electricity *obstacles* *spillages*	*contractors* *maintenance*

Keep a record of your findings

If quite a lot of improvements need to be made, it's best not to try and do everything at once, but to draw up an action plan to deal with the most important ones first.

A good action plan will often include implementing a few cheap, or easy, improvements which can be done quickly, perhaps as a temporary solution, until more reliable controls can be put in place.

The plan of action might also include long-term solutions for those risks which are most likely to cause accidents or ill-health, or which have the worst potential consequences, as well as the arrangements which are made for training employees on how the remaining risks will be controlled.

And finally, the action plan should include who has responsibility for the various actions and by when, as well as how regular checks will be made, to make sure that control measures stay in place.

Date of risk assessment:		
Step 3 What are you already doing?	What further action is necessary?	**Step 4** How will you put the assessment into action?
List what is already in place to reduce the likelihood of harm or make any harm less serious.	You need to make sure that you have reduced risks "so far as is reasonably practicable". An easy way of doing this is to compare what you are already doing with good practice. If there is a difference, list what needs to be done.	Remember to prioritise. Deal with those hazards that are high-risk and have serious consequences first. Action Action Done by whom by when

Draw up an action plan

Step 5 – Review risk assessment and update if necessary

Having identified the hazards, who might be harmed by them, and how, worked out what the risks are and the controls that are needed, and kept a record of what's been done, the fifth step in the risk assessment process is to review the risk controls, and to update them as necessary.

It's a good idea to do this on an on-going basis, by thinking about risk assessment when changes are being planned, as well as conducting a formal, annual review.

Reviewing the risk assessment regularly will mean that controls can be amended each time new hazards are introduced, for example, when there are significant changes in the workplace, or with the introduction of new equipment, substances, or procedures, when problems are spotted by employees, or when accidents or near misses occur.

By carrying out regular reviews, you can ensure that risk controls are always up to date and improving, and not sliding back.

Fifth Step is to review risk assessment and update if necessary

Responsibilities

Both employers and employees have responsibilities for health and safety in the workplace.

Employers are responsible for ensuring risk assessments are carried out on a regular basis. The process doesn't need to be a complicated one, nor does it need a health and safety expert to do one.

Employees have a responsibility to co-operate with their employer's efforts to improve health and safety, by complying with the controls which are in place, and by looking out for each other.

Risk assessment form

FIRE PROTECTION

Classes of fire

There are four common classes of fire:

- Class A: Solids – wood, paper, textiles, etc.
- Class B: Flammable liquids – oil, petrol, paint, etc.
- Class C: Flammable gases – acetylene, propane, butane, etc.
- Class D: Metals – magnesium, aluminium, sodium, etc.

Class A

Class B

Class C

Class D

The type of burning material tells you which type of fire extinguisher to use.

As electrical fires do not fall into any particular 'class' of fire, if an electric spark ignites, for example, paper, you would use a Class A extinguisher.

Types of fire extinguisher

There are four different types of fire extinguisher, which are shown.

A water fire extinguisher can be used for class A fires

Foam fire extinguishers can be used for class B & D fires

Powder fire extinguishers can be used for all classes of fire: A, B, C and D

A carbon dioxide fire extinguisher can be used for class B, C and D fires

All fire extinguishers are now red, and are labelled to identify which type is which. It's important to select the correct one for the class of fire, otherwise it could have serious consequences.

- A water fire extinguisher is suitable for class A fires
- A carbon dioxide fire extinguisher is suitable for class B, C and D fires
- A foam fire extinguisher is suitable for class B and D fires
- A powder fire extinguisher is suitable for all classes of fire (A, B, C and D)

CSCS SITE SAFETY CARDS

There are nine different types of site safety card in the Construction Skills Certification Scheme, or **CSCS** as shown.

CSCS Construction Skills Certification Scheme

CSCS cards

Each card is issued to people in relation to their relevant experience and qualifications, and requires completion of the appropriate health and safety course or test.

Here are the details of the types of people who qualify for each of the two red cards.

Red Card – Trainee (Craft and Operative): registered for NVQ or SVQ (or Construction Award) and not yet reached level 2 or 3.

Red Card – Trainee (Technical, Supervisory and Management): registered with a further/higher education college for a nationally-recognized, construction-related qualification or satisfactorily completed such a course.

Here are the details of who qualifies for a green card.

Green Card – Construction Site Operative: who carries out basic site skills with Level 1 NVQ or employer's recommendation using industry accreditation.

Here are the details of the types of people who qualify for each of the three blue cards.

Blue Card – Experienced Worker: with at least one year's on-the-job experience in last three years who missed opportunity for industry accreditation. Card is valid for one year whilst achieving Level 2 or higher NVQ/SVQ but is not renewable.

Blue Card – Skilled Worker: with Level 2 or higher NVQ/SVQ or completed employer sponsored apprenticeship, or completed City & Guilds of London Institute Craft Certificate.

Blue Card – Experienced Manager: with at least one year's on-the-job experience in last three years. Card is valid for three years whilst achieving Level 3 or higher NVQ/SVQ but is not renewable.

Here are the details of who qualifies for a gold card.

Gold Card – Skilled Worker: with Level 3 NVQ/SVQ or approved indentured apprenticeship or employer sponsored apprenticeship and completed City & Guilds of London Institute Advanced Craft Certificate.

Gold Card – Supervisor: with Level 3 NVQ/SVQ in supervisory occupation or industry accredited.

Here are the qualifications required for a platinum card.

Platinum Card – Manager: with industry accreditation or Level 4 NVQ/SVQ.

Here are the qualifications required to obtain a black card.

Black Card – Senior Manager: with Level 5 NVQ/SVQ.

Here are the details of the type of people who qualify for a yellow card.

Yellow Card – Professionally Qualified Person: consultants who are chartered members of approved institutions (e.g. architects, surveyors and engineers) with health and safety responsibilities and on site no more than 30 days in six month period.

Here are the details of who is eligible for a white card

White Card – Construction Related Occupation: for those occupations not covered by other cards.

Here are the details of the yellow visitor card.

Yellow Card – Regular Visitor: with no specific construction skills who often visits a construction site. Makes site access easier as the holder would have passed a health and safety test before the card was issued.

ELECTRICAL SAFETY

Looking after and maintaining electrical equipment

The bottom line is that all electrical equipment should be looked after, and maintained in a safe condition. But this doesn't always happen, as these pictures show.

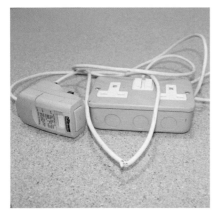

Extension leads that are moved a great deal are prone to damage

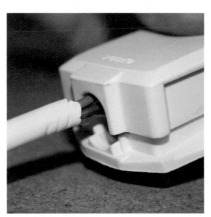

Plug cables should be firmly clamped

Always use proper connectors to join cables

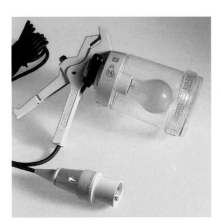

Equipment that can easily be damaged must be protected

Suspect or faulty electrical equipment must be labelled "DO NOT USE"

Some equipment is not suitable for use in wet or harsh environments

- Extension leads that are moved a great deal are particularly prone to being damaged. If the cable, plug or socket is damaged it should be replaced.

- The outer sheath of flexible cables must always be firmly clamped to stop the wires (particularly the earth) from pulling out of the terminals.

- Cables should always be joined with proper connectors or cable couplers, not with strip connectors and insulating tape.

- Lamps and equipment which can easily be damaged must be protected to prevent risk of electric shock.

- Suspect or faulty electrical equipment must be labelled 'DO NOT USE' and kept secure until it can be examined by a competent person.

- Equipment unsuited for use in a wet or harsh environment can easily become live and also make the surroundings live.

Pre-visual inspection

Many faults with electrically operated power tools can be found by visual inspection, but by following a simple process before using the equipment, you can minimize most electrical risks, as shown here.

- Switch off and unplug

- Check plug is correctly wired

- Check fuse is correctly rated by checking equipment rating plate or instruction book

- Check plug is not damaged, cable is properly secured and no internal wires are visible

- Check cable is not damaged and has not been repaired with insulating tape or unsuitable connector

- Check outer cover of equipment is not damaged which might give rise to electrical or mechanical hazards

- Check equipment for burn marks or staining that might suggest equipment is overheating

Switch off

Unplug the equipment

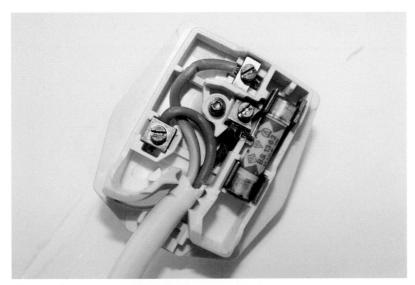

Check plug is correctly wired

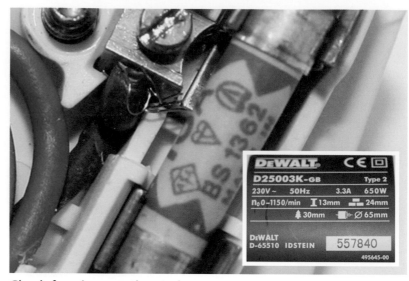

Check fuse is correctly rated

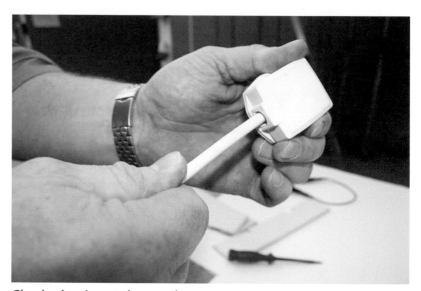

Check plug is not damaged

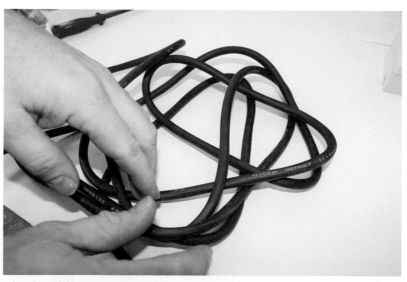

Check cable is not damaged

Check outer cover of equipment is not damaged

Check equipment for burn marks

E-LEARNING

Use the e-learning programme to see a demonstration of a pre-visual inspection.

Electrical site safety

UK AND INTERNATIONAL STANDARDS

When it comes to the safe isolation of electrical supplies and energizing electrical installations, it is also important to comply with statutory health and safety requirements, as laid down by the Electricity at Work Regulations.

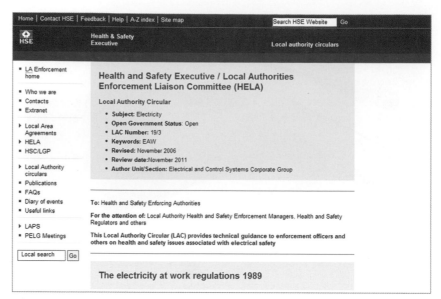

HSE Website

Safe isolation of electrical supplies

In order to avoid fatal accidents, which can occur during the proving of isolation, you should follow the recognized procedure:

- Identify source of supply
- Identify type of supply
- Isolate
- Secure the isolation
- Test the equipment/system is dead
- Begin work

Identify the source of supply

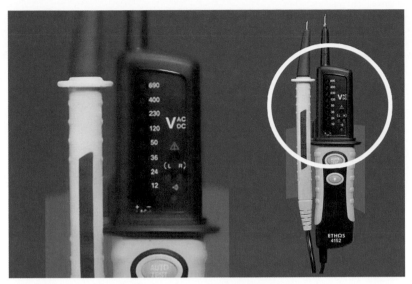

Identify type of supply

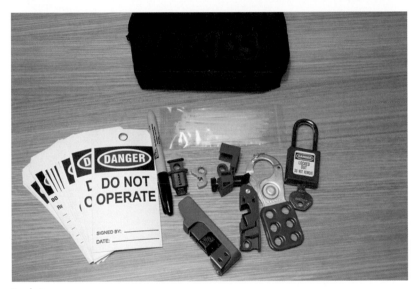

Isolate

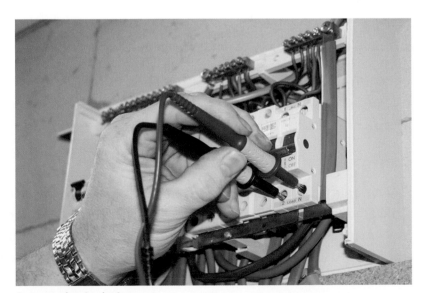

Secure the isolation

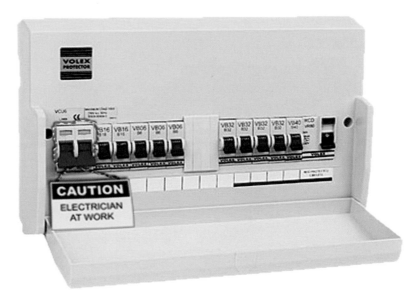

Test the equipment/system is dead

Begin work

E-LEARNING

Use the e-learning programme to see a demonstration of safe isolation.

Energizing electrical installations

A number of deaths and major injuries have occurred when electrical circuits have been energized at the request of building designers, clients, contractors, or finishing trades before the electrical installation was complete. It is not considered 'reasonable to work live' solely on the grounds of inconvenience, lost time, or cost.

Electrical contractors are only able to energize circuits when it is unreasonable to work dead, and a written request has been made by the main contractor or his agent. Suitable precautions and testing must also be undertaken, before the electrical contractor agrees it is safe to energize the circuit.

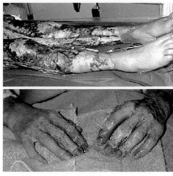

Many major injuries have occurred when electrical circuits have been energized before installation was complete

Other aspects of electrical site safety

Here are some more ways in which you can ensure site safety with regard to electricity.

Use reduced voltage equipment to reduce risk of injury

Use battery-operated power tools - they are safest to use

Portable tools are readily available

Using Residual Current Device (RCD) can reduce injury

If working is near overhead power lines they should be switched off

CHECK YOUR KNOWLEDGE

1. Imagine you have returned to a job with only a few tasks remaining. You have been wearing overalls and a hard hat, but your gear is in the van. The risks don't seem serious, and the work will only take a few minutes. Do you have to use your PPE?

    ☐ a. No – the work will only take a few minutes

    ☐ b. No – the risks do not seem serious

    ☐ c. Yes – the PPE should still be used

2. Under the Control of Substances Hazardous to Health Regulations – and European law – who has overall responsibility for controlling exposure to hazardous substances in the workplace?

    ☐ a. Employers

    ☐ b. Employees

    ☐ c. Government Health and Safety Inspectors

3. What items do you think should be part of a first aid kit?
    Select 7 items and add them to the table shown.

Item

4. Here are the five steps in the risk assessment process. Put the
    steps in the right order in the table shown.

    ☐ a. Who might be harmed and how?

    ☐ b. Review risk assessment and update if necessary

    ☐ c. Record findings and implement them

    ☐ d. Identify the hazards

    ☐ e. Evaluate the risks and decide on precautions

Step	Description
1.	
2.	
3.	
4.	
5.	

5. **Which *one* of the following fire extinguishers is suitable for use on all classes of fire?**

   ☐ a. Carbon dioxide

   ☐ b. Foam

   ☐ c. Powder

   ☐ d. Water

6. **If you were visually inspecting this piece of equipment how many faults would you find?**

   ☐ a. One

   ☐ b. Two

   ☐ c. Three

   ☐ d. Four

7. **Ladders and scaffolding are the means most installers use to enable them to work at heights. However, they are not always used safely. What percentage of accidents reported to the Health and Safety Executive do you think are due to falls by people working at heights?**

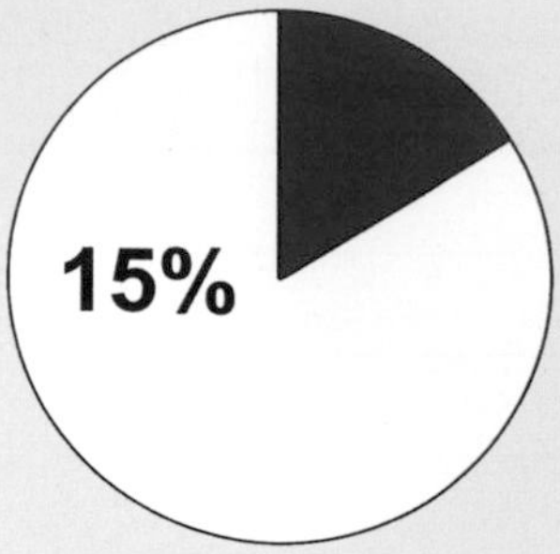

15%

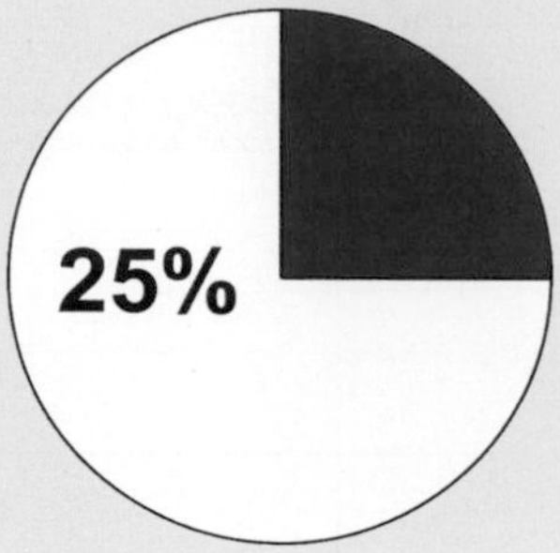

25%

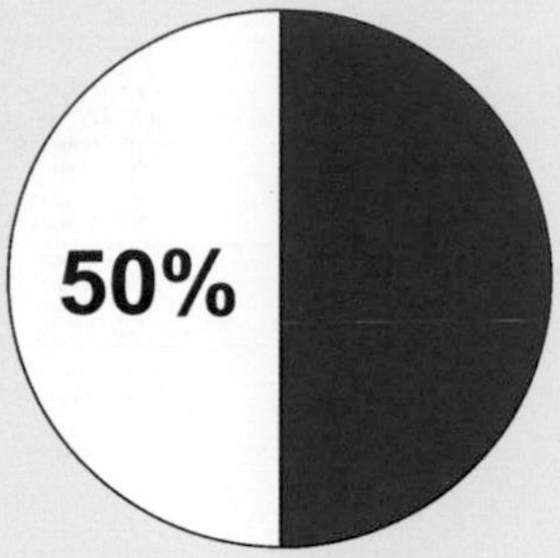

50%

8. Match up the fire extinguishers shown with the class of fire they are used on.

Class of Fire	Fire Extinguisher
Class A – wood, paper, textiles, etc	
Class B – oil, petrol, paint, etc	
Class C – gas, acetylene, butane, etc	
Class D–metal, magnesium, aluminium, etc	

Chapter 3

SOLAR COLLECTOR TYPES AND UK AND INTERNATIONAL STANDARDS

LEARNING OBJECTIVES

By the end of this chapter you will be able to:

- Identify the UK and International Standards relevant to solar hot water systems

- Describe how a solar collector absorbs energy

- Describe different types of collectors that can be used for solar hot water systems

- Identify appropriate types of roof mounting for different types of collectors

UK AND INTERNATIONAL STANDARDS FOR SOLAR COLLECTORS

British Standard The British Standards Institution (BSI) sets quality standards and standard dimensions for equipment and materials. All British Standards start with the letters BS followed by a number. BS EN 12975:2006 Thermal solar systems and components – solar collectors is in two parts, durability and performance of the solar collector. BS5918, code of practice for solar heating, is also relevant.

All solar collectors need to be durable, maximize absorption of solar energy and limit energy losses. The **British Standard** to remember is BS EN 12975:2006, 'Thermal solar systems and components – solar collectors' and it is in two parts. It addresses questions about durability and performance of the solar collector. As this is an EN standard, it is a pan-European standard that applies across the whole continent. For example in Germany, this same standard is called DIN EN 12975:2006 and is written in German rather than English.

There are various other British Standards relevant to solar thermal such as BS 5918, code of practice for solar heating. Here, we are only looking at the solar collector standards. BS 5918 is a standard that is local to the UK. Similar installation standards are frequently used in various different European countries and various authorities are currently working in Brussels to, as far as possible, harmonize these standards.

EN 12975:2006 - Thermal solar systems and components – solar collectors

The Standard is in two parts. The two parts describe durability testing and performance testing.

Together they address the main requirements of any collector, which are to:

● Be robust and durable

EN 12975: 2006

- Maximize absorption of solar radiation
- Limit energy losses

Other solar standards are not discussed in this chapter.

EN 12975 durability

The function of a solar collector is to absorb solar radiation for heating a transfer fluid, therefore it needs to last a long time. EN 12975 tests solar collectors to a durability standard. This looks at the issues relating to mechanical strength, high temperatures both inside and outside the collector, factors affected by weather, and possible problems with its position on the roof. The collector must pass these tests of mechanical strength, reliability and safety. Please note that you are not expected to learn the full list of durability tests! The list is included for your reference.

EN 12975 durability tests

- Resistance to leakage and distortion from internal pressure
- Ability to withstand high temperature without fluid
- Ability to withstand long periods without fluid (exposure)
- An internal temperature shock test
- An external temperature shock test
- Resistance to rain penetration
- Ability to resist frost
- A downward pressure test on the **glazing** simulating wind and snow loads
- An upward pressure on the glazing simulating wind lift
- An upward pressure on the collector fixing brackets
- Ability of the glazing to withstand impact

Glazing Glazing materials for the collector need to let the maximum solar energy through to the absorber and the minimum to be transferred back to the atmosphere

The EN 12975 durability tests

EN 12975 performance

EN 12975 also measures performance. This measurement test records the energy output of a solar collector over a number of days. A minimum solar radiation level must be reached during the day to qualify as a test day. This test records a level of performance, rather than a pass or fail of the individual test, as with the durability tests.

The EN 12975 standard also states that collectors should be labelled with specific information and have an installer instruction manual.

EN 12975 performance test

EN 12975 performance test

- The performance test records energy output of a solar collector over a number of days when solar radiation is above a minimum value.

- The performance test records the efficiency of the collector.

- (The durability test provides pass or fail criteria for the collector.)

EN 12975 also specifies the items that must be included on a label fixed to the collector, and that an installation instruction manual should also be included with the collector.

ACTIVITY 7

If you were asked to go down to the local hardware store and buy some materials and build a solar collector, what might you buy and do you think it is likely to pass the EN 12975 durability and performance test? (Clue: you might want to study some of the information on flat plate collectors below before answering this activity. That is unless you are building evacuated tube collectors in your garden shed! In which case, you will need a vacuum cleaner to create the vacuum.)

Flat plate collector Flat plate collectors are glazed and insulated and these are used for solar hot water heating systems. Unglazed flat plate collectors are often used to heat swimming pools but not for hot water systems

Evacuated tube collector Evacuated tube collectors are made up of a series of evacuated (empty) tubes connected to a manifold pipe assembly, which is all housed in a well insulated metal box. Evacuated tube collectors are nearly always installed on top of roof tiles

TYPES OF SOLAR COLLECTOR

Types of collector

The main component of a solar heating system is the collector. Its function is to collect the energy falling on it and transfer it in the form of heat to the fluid in the collector.

There are several types of solar collectors, the main two being **flat plate collectors** and **evacuated tube collectors**. We shall look at each in turn.

Flat plate collector   *Evacuated tube collectors*

Flat plate collectors

The flat plate collector is probably the most common solar panel, and it is basically a solar energy absorber fitted into an insulated box with a glazed surface facing the Sun. The box forms a weatherproof casing to protect the absorber plate and its insulation. Flat plate collectors are glazed and insulated and these are used for solar hot water heating systems. Unglazed flat plate collectors are often used to heat swimming pools but not for hot water systems. Flat plate collectors:

- Have insulation at the back and sides of the weatherproof box
- Have a glazed surface facing the Sun
- Have an absorber plate made up of a series of pipes attached to one or more metal fins spanning almost the full surface area of the collector.

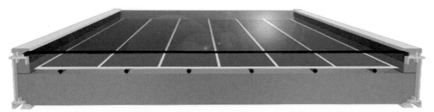

Flat plate collectors are usually the most common solar panels used

Glazing

The glazing should allow the maximum amount of solar radiation to pass through to the absorber. It should also prevent the heat being transferred back into the atmosphere and keep this heat loss to a minimum.

Glazing materials can be made from:

- Glass: which has good transmission rates but is heavier than acrylic or plastic film. High quality glass transmission rates are between 91 per cent and 96 per cent

- Acrylic: which is blow-formed to increase rigidity and has a transmission rate of 89 per cent

- Plastic film: which has a high UV resistance and is very light weight. It has transmission values of 90–95 per cent with Tedlar or Teflon coating. However, it needs protection from birds.

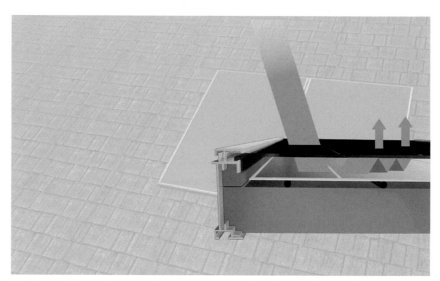

Glazing allows the maximum amount of radiation in and prevents heat from transferring back into the atmosphere

The casing

The collector casing can be assembled in the factory and can be made of aluminium, wood or steel. The box corners are sealed to prevent rain getting in, however there are often protected ventilation holes to allow condensation to escape and release air pressure in the box if temperatures get very high. The insulation, such as rock wool, must be capable of withstanding temperatures of up to 210°C and must neither melt nor give off vapours at these temperatures.

E-LEARNING

Use the e-learning programme to learn more about the glazing and casing.

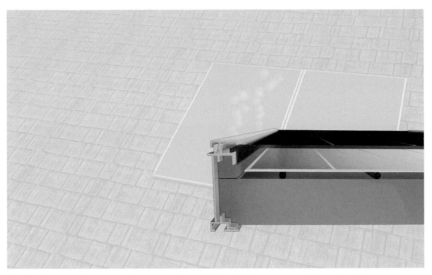

Collector casing can be made of aluminium, wood or steel

Absorber plate

The absorber is sometimes plastic. However, the significant majority of absorber plates are made from copper or aluminium, with copper pipes brazed or welded to the absorber plate. A liquid (or air in the case of a ventilation-based collector) is passed through the pipes to extract the heat. The Sun-facing surface of the absorber plate is normally coated with a black 'selective surface' which is specially designed to absorb as much solar energy as possible without letting go of this solar energy as infrared radiation. This is why it is called a selective surface, as it selects the Sun's energy without losing much of this energy as infrared. A black painted surface also works but it does not retain as much energy as a selective surface. Similar selective surface absorber plates are also used in evacuated tubes. Selective surfaces are further discussed later on in this chapter.

Flat plate collector summary

The information on flat plate collectors is summarized in the list below, but with some additional information about stagnation temperatures and positioning of the collectors on buildings.

- Unit size range from $1m^2$ to $4.5m^2$
- Has insulation to limit heat loss from the collector
- Glazing material can be toughened glass, acrylic, polycarbonate or Teflon/Tedlar coatings
- Can be installed on top of roof tiles or integrated into the roof covering

- Stagnation temperature range 170°C to 210°C
- Versatile mounting options including low roof angle and integrated vertical facade installations
- Generally more acceptable for planning applications in conservation areas and Areas of Outstanding Natural Beauty (AONB)

Flat plate collector

Evacuated tube collector

An evacuated tube collector is a series of evacuated tubes connected to a manifold pipe assembly at the top, which is housed in a well-insulated metal box. The vacuum inside the tube acts as a very efficient heat insulator. Each tube consists of a double or single wall glass outer tube containing a flat or cylindrical absorber. A pipe runs inside the full length of the tube, which carries the heat transfer fluid to the top of the tube where it meets the manifold.

Evacuated tube collectors are nearly always installed over roof tiles as the gaps between the tubes make roof integration impractical.

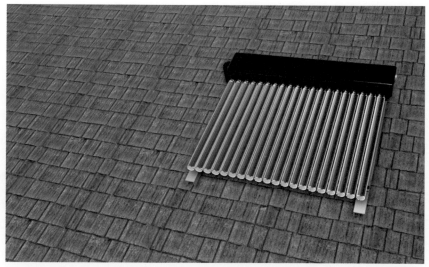

Evacuated tube collector - the vacuum inside acts a heat insulator

There are two types of evacuated tube collector, which work in different ways; direct flow and heat pipe collectors. Both types can potentially deliver higher energy outputs compared to flat plate collectors of the same total **absorber surface area**. This is due to their lower heat loss. Government sponsored test reports suggest that annual benefits can be in the region of 15 per cent to 30 per cent per m^2. However, tube collectors are generally more expensive and the difference can often be made up by increasing the surface area of flat plate collectors.

Both are more efficient than flat plate collectors for the same absorber surface area.

Evacuated tubes are useful when the roof surface area is restricted.

> **Absorber surface area** The surface on the collector where solar energy is absorbed

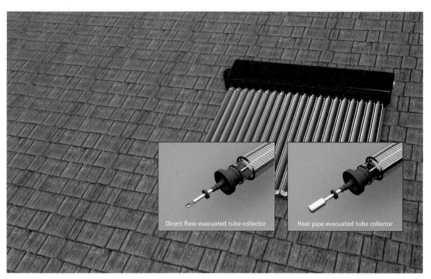

Direct flow evacuated tube collector   Heat pipe evacuated tube collector

Two types of evacuated tube collector; direct flow and heat pipe collectors

Direct flow evacuated tube collectors

Direct flow evacuated tube collectors are called 'Direct Flow' because the heat transfer liquid passes through the manifold pipe and each evacuated tube in the assembly. They have a high stagnation temperature of 250°C to 300°C. They can be built into arrays of between 6 and 30 tubes in one single manifold collector and have versatile mounting options, from vertical to horizontal, portrait and landscape. Individual tube diameters can vary between 80mm and 250mm.

> **Direct flow evacuated tube collector** Type of solar panel. The heat transfer liquid passes through the manifold pipe and each of the evacuated tubes in the assembly

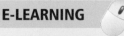

E-LEARNING

Use the e-learning to see a demonstration of Direct flow evacuated tube collectors.

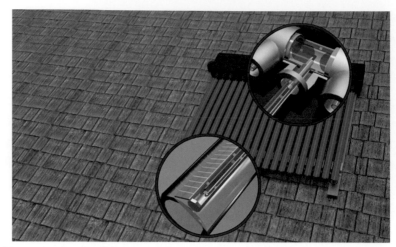

The transfer liquid passes through the manifold pipe and each tube

Heat pipe evacuated tube collectors

Heat pipes work in a completely different way. Each tube is a separate sealed pipe with a large heat transfer 'plug' at the top. The heat transfer fluid within the tube is also sealed in and does not mix with the fluid circulating in the manifold. The heat transfer fluid in the tube boils, evaporates and rises within the tube, thus releasing a lot of heat energy at the condensing tip which is transferred to the passing fluid in the manifold. The condensed liquid then trickles down the evacuated tube ready to be heated again. It then condenses on the copper 'plug' – also known as the 'condensing tip', where it loses its heat through conduction to the cooler fluid passing by in the manifold pipe.

- Available in arrays of between 20 and 30 tubes connected to a single top manifold assembly

- Stagnation temperature range 220°C to 250°C

- Normally have a minimum mounting angle (around 25° to horizontal) and must never be mounted horizontally as they work by gravity

> Heat pipe evacuated tube collectors Each pipe is a sealed unit with a large heat transfer 'plug' at the top

Each tube is a separate sealed pipe with large heat transfer plug

ACTIVITY 8

Evacuated tubes are normally a round shape. Why is this shape particularly useful and what would probably happen if the tubes were made square with thin glass?

What is the main advantage of direct flow tubes as compared to heat pipe tubes and vice versa. What is the main advantage of heat pipe tubes as compared to direct flow evacuated tubes?

Unglazed and uninsulated collectors

Where the required heat is only just above ambient air temperature, such as heating for indoor or outdoor swimming pools, there is little need for insulation. Radiation and convection losses from solar collectors are proportional to the difference between the collector fluid and ambient, or air temperature. When these are small it is easier to design for maximum heat gain rather than minimize heat loss. The collector just needs to be in a sheltered position where wind speed is minimized. These collectors can be made of plastic panels, strips or pipes.

If ambient air temperature required - unglazed and uninsulated collectors are ideal

HOW SOLAR COLLECTORS ABSORB ENERGY

Function of the absorber

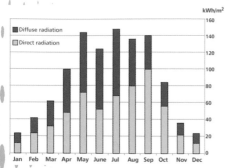

Diffused and direct radiation throughout the year

Direct and diffuse solar radiation reaches the UK all year round, but due to our latitude it varies throughout the year. On a flat surface, most of the UK receives between 900kWh and 1100kWh per m^2 per year and it is spread through the year as in the graph. The absorber needs to capture this energy and turn it into useful heat.

There are three definitions regarding the surface area of a solar collector:

- Total surface area – includes casing
- Aperture surface area – through which light can enter
- Absorber surface area – where solar energy is absorbed

The absorber itself usually consists of a surface, usually flat, on which the solar radiation falls. Tubes or channels are either attached to it or formed in it, in which the heat transfer fluid circulates collecting heat from the surface.

Materials used for absorbers are frequently copper or aluminium with copper tubes, and the absorber must be resistant to attack by the heat transfer fluid. Absorbers must also be constructed to resist the very high temperatures during periods of stagnation.

UK AND INTERNATIONAL STANDARDS

Various claims are made as to the benefits of flat and round absorbers. The recommended way to check collector performance is to study and compare the EN 12975 performance test.

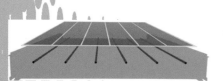

Absorber - flat surface where solar radiation falls

ACTIVITY 9

On the diagram of a flat plate collector, label the total surface area, aperture surface area and absorber surface area.

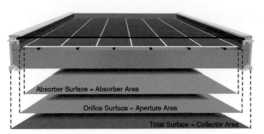

Absorber Surface = Absorber Area
Orifice Surface = Aperture Area
Total Surface = Collector Area

Three definitions regarding the surface area of a solar collector

Reflectance, absorptance, radiance

Solar radiation reaching a collector is absorbed subject to three factors:

- Reflectance
- Absorptance
- Radiance

The first two need to be maximized and the last one minimized.

Reflectance is the fraction of light deflected away when it strikes the glazing of a collector.

Absorptance is the amount of solar radiation being absorbed by a solar collector, typically more than 90 per cent

Radiance is the amount of absorbed energy radiated back out of the collector, typically between 4 per cent and 12 per cent

Reflectance This is the fraction of light deflected away when it strikes the glazing of a collector

Absorptance This is the amount of solar radiation being absorbed by a solar collector, typically more than 90 per cent

Radiance This is the amount of absorbed energy radiated back out of the collector, typically between 4 per cent and 12 per cent

E-LEARNING

Use the e-learning programme to see a demonstration of Reflectance, Absorptance and Radiance.

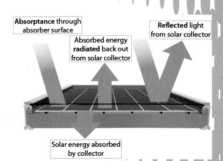

Absorptance through absorber surface
Absorbed energy radiated back out from solar collector
Reflected light from solar collector
Solar energy absorbed by collector

3 factors; reflectance, absorptance and radiance

Selective and non-selective surfaces

The absorber needs to capture as much radiation as possible and that includes diffuse radiation as well as direct sunlight. Dark colours and matt surfaces absorb more radiation than light colours and polished surfaces. Absorbers are therefore coated to increase their capacity to absorb solar radiation. The coating can be selective or non-selective.

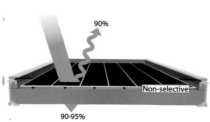

Non-selective

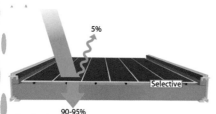

Selective

A typical non-selective surface would be black paint, which improves absorbance of solar radiation, but also increases radiation emitted from the absorber as the temperature rises. Non-selective coatings are good absorbers of solar radiation (90 per cent to 95 per cent) but lose about 90 per cent by radiating it back out from the absorber. The efficiency of the collector falls rapidly as the temperature of the heat transfer fluid rises.

A selective surface would increase absorbance, but not radiate back as much energy, therefore a selective coating increases the efficiency of the absorber. Selective coatings are good absorbers of solar radiation (90 per cent to 95 per cent) and only lose between 4 per cent and 12 per cent of the energy through radiation from the absorber. This makes the efficiency of the selective surface collectors much greater than simple black paint collectors.

ACTIVITY 10

An absorber surface is said to have Reflectance, Absorptance and Radiance in the way it handles (the physics of) light. The glazing of the collector is also an important part of the light physics of a solar collector. The glazing will also reflect, absorb (normally only a very small amount) and radiate light. What other physics of light feature will happen in the transparent glazing cover of the solar collector? What efficiency effect does this have on the solar collector?

COLLECTOR EFFICIENCY

Collector efficiency is expressed as a ratio, or percentage, comparing the amount of heat extracted from the collector by the heat transfer fluid with the total solar energy falling on the collector. The collector efficiency is measured for the collector as a whole, and not just the absorber or aperture area. The difference in temperature between the air or ambient temperature, and the temperature of the collector, is very important when considering collector efficiency and is known as Delta T (ΔT).

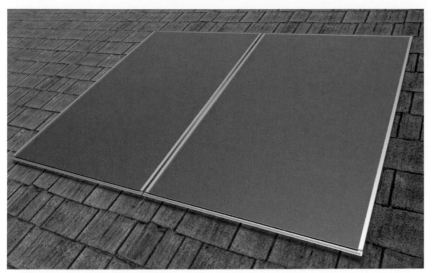

Collector efficiency compares the amount of heat extracted from the collector

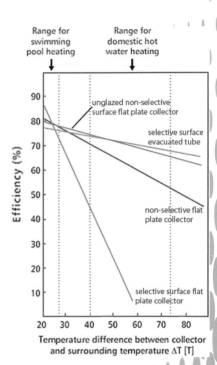

Efficiency comparison at irradiation of 1000 W/m²k

FUNCTIONAL SKILLS

The graph shows the temperature difference between four types of collectors and their efficiency at a range of collector temperatures above ambient temperature, the ΔT measurement. We have a selective surface evacuated tube; a selective surface flat plate collector; a non-selective flat plate collector and an unglazed non-selective surface flat plate collector.

At low temperatures the efficiency of all collector types is similar, but as the temperature difference rises the efficiency decreases. Unglazed non-selective surface flat plate collectors are most efficient at lower temperatures and therefore very efficient for heating swimming pools, but not as good for a domestic hot water system.

The difference between selective surfaces on flat plate and tube collectors is not very significant over the temperature range for hot water supply, with evacuated tube collectors being the least sensitive to increases in temperature difference. It can be seen that all glazed collectors can raise the temperature to those required by a domestic hot water system, and as efficiency can be compensated for by increasing the solar panel area, domestic systems can be built to suit the needs of the customer and still provide a useful resource. Nearly all modern collectors have a selective surface to gain these efficiency benefits.

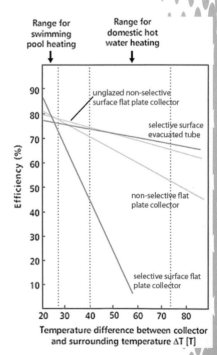

ACTIVITY 11

Looking at the previous graphs, what is the efficiency of the four different collector types at 20 ΔT and 54 ΔT?

Energy outputs

Most solar energy falls on the UK during the summer, with significant inputs during spring and autumn. The efficiency of the solar collector is important throughout the whole year, as this affects how much contribution can be made to the domestic hot water requirements of a household. Solar panels under clear sky conditions in late spring and early autumn can still boil water and reach high collector surface temperatures.

During the year it is highly unlikely solar radiation can supply all the domestic hot water requirements:

- During summer solar collectors can supply up to 100 per cent of domestic hot water requirements

- During winter solar collectors still provide preheating for hot water systems and an alternative heating supply (e.g. gas or electric) can back up hot water needs

- In spring and autumn contribution from solar radiation gradually increases/decreases as the seasons change

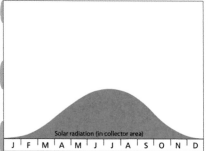

Most of the solar energy in the UK falls during the summer

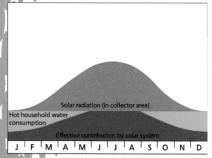

In the summer months solar collectors can supply 100% of hot water requirements

ACTIVITY 12

You are specifying a solar system to:
1. Heat just a swimming pool for summer use only
2. Heat hot water and a swimming pool
3. Heat hot water and provide some of the space heating load in spring and autumn
   Please state what collector choice you might consider.

MOUNTING COLLECTORS ON ROOFS

HEALTH AND SAFETY

The most important thing to remember when considering fitting solar collectors to roofs is to make sure they are very firmly fixed on to that roof to prevent both wind lift from removing them and that they are weatherproof. The manufacturers' instructions must be followed at all times to ensure everyone's safety.

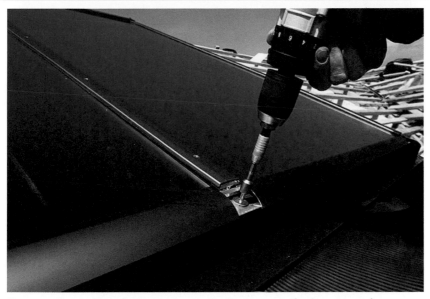

Ensure that when fitting solar collectors to roofs they must be firmly fixed

Collector mounting arrangements

When collectors are mounted on roofs it cannot be emphasized too strongly that adequate fixing of the solar collectors against wind lift and rain penetration is essential. Manufacturers' instructions must be adhered to at all times.

There are various ways solar collectors can be positioned, and this usually means either on a frame on a sloping roof, or integrated within the roof similar to a roof light window. They can also be mounted on a frame on a wall or at ground level. They can be fixed to a frame on a flat roof, but as flat roofs have a limited life you need to make sure the roof will last as long as the collectors!

UK AND INTERNATIONAL STANDARDS

Some arrangements that stick up more than 200mm above the roof line may need planning permission (depending on the local planning regulations).

You should always fit the frame before lifting the collector to the roof, as it reduces the time the collector is on the roof unsecured.

Adequate fixing of solar collectors against rain/wind lift is essential

They can be positioned on a sloping roof or integrated within the roof

They can be fixed to a frame on a flat roof or in a garden

Roof penetration methods for solar pipes from on-roof collectors

The pipes from the collectors on the roof need access to the rest of the solar domestic hot water system inside the house. There are a range of methods, three are shown here.

The first uses lead flashing containing a silicone rubber bonnet (sometimes called a solar **dektite**) where the holes are cut for the pipes.

The second method shows a custom made roof tile, which can be bought at a roof suppliers.

The third uses a roof tile with a hole drilled in it and the pipes are secured using a silicone sealant. This third method does not qualify for certification criteria.

Dektite Lead flashing containing a silicone rubber bonnet – holes are cut in the bonnet for the pipes to pass through

First method

Second method

Third method

In-roof mounting methods

Flat plate panels can be integrated into the roof and the arrangement is similar to integrated roof windows with flashings providing a waterproof seal. This method is often done with a new build or when a house is completely re-roofed although it is also possible to do as a retrofit. Measurements must be carefully made and it is important to check the associated pipework does not coincide with rafters and trusses. Collectors can be made flush with the roofline.

Flat plate panels can be integrated into the roof with flashings providing a waterproof seal

ACTIVITY 13

When mounting collectors on domestic roofs, three types of roof are typically used:
1. Single lapped flat or profiled concrete tiles
2. Double lapped plain tiles
3. Double lapped slate roofs

Of course there are many other roofing materials and options such as metal or corrugated roofs. However, on domestic solar, most tiles and slates can be fitted into these three types of roofing material.

Thinking about each type of roof tile or slate, discuss the easiest options for retrofitting a collector both in-roof and on-roof. It might be helpful to briefly Google the different roofing options and also to briefly look at how roof window flashing kits are designed.

CHECK YOUR KNOWLEDGE

1. **Which of the three types of solar collector shown <u>does not</u> contain transfer fluid at point X? Tick the right image.**

Direct flow tube collector

Heat pipe tube collector

Flat panel collector

2. **Solar collectors must perform up to British Standards (BS EN 12975). The standard is split into two parts; Durability Testing and Performance Testing.**

Place each of the statements listed here into the relevant category.

☐ a. Ability to resist leakage

☐ b. Ability to record efficiency

☐ c. Ability to resist distortion from internal pressure

☐ d. Ability to withstand high temperature without fluid

☐ e. Ability to resist frost

☐ f. Ability to resist rain penetration

☐ g. Ability to record energy output

Durability Testing	Performance Testing

3. **A value △T (Delta T) is used when measuring the efficiency of solar collectors. From which two measurements is this △T calculated?**

☐ a. Temperature of transfer fluid before and after heating in the collector

☐ b. Ambient temperature and collector temperature

☐ c. Ambient temperature and transfer fluid temperature before it enters the collector

☐ d. Temperature of the collector surface and the absorber surface

4. **A heat pipe evacuated tube collector must be mounted with a minimum mounting angle. Draw a heat pipe evacuated tube collector showing this angle from the horizontal.**

Chapter 4

SOLAR THERMAL STORE OPTIONS

LEARNING OBJECTIVES

By the end of this chapter you will be able to:

- Identify the purpose of a storage vessel within a solar hot water system

- Describe the types of solar hot water system storage vessels

- List considerations required if asked to install solar hot water into a system with an instantaneous water heater (combi boiler)

- Describe solar storage vessel hazards

Options for solar thermal storing

SOLAR STORAGE VESSELS

Storage vessel requirements

Solar domestic hot water systems absorb energy during the day, but people use hot water at any time of the day and night and therefore a means of storing the hot water is needed. As the supply of solar energy in the UK is not enough to provide all our hot water requirements, the hot water storage also needs to have another source of energy, usually from a conventional gas, oil, electric or other renewable heat source.

The hot water storage needs to have another source of energy

Basic system plans and shapes

Storage vessels used in solar hot water systems can be of a variety of designs but all require a cold supply, directly fed in the case of an unvented hot water cylinder, or via gravity from a cold water cistern; a suitably sized storage volume for the solar energy contribution; a means of receiving the solar energy, for example a heat exchange coil; a suitably sized storage volume for hot water from the back up heat source; a means of receiving the back up energy (for example, an immersion heater) and a hot water outlet. The storage method can be a single cylinder with two heat supplies or two cylinders, each with its own heat source.

Storage vessels used in hot water systems all require a cold supply

Solar hot water storage vessels can also benefit from a few other changes. Cylinders that are taller and thinner allow stratification of the water, with the hottest at the top and the coolest at the bottom. The cooler the solar section of the cylinder dedicated to solar heating, the better the system's efficiency. Extra insulation helps both the solar and back up systems be more efficient. Making sure that the volume of water to be heated by the solar circuit is correctly sized makes a big difference – if it is too big the total volume is cooler for the same energy input, and if the water volume is too small, it will overheat and cause stagnation to occur more often. Well placed sensor pockets give better control of the system and increase the solar contribution.

Things that can add benefits:

- A taller, thinner shape helps to separate hot and cold water
- Extra insulation, both on cylinder walls and pipework leading to and from the cylinder
- Efficient sizing of storage volume to solar energy, at least 80 per cent or more of daily hot water demand improves efficiency
- too big – lots of lukewarm rather than hot water
- too small – overheating and regular stagnation
- Well located sensor pockets for control of solar energy contribution

Cylinders that are taller and thinner allow stratification of water

There is much discussion and debate as to when stratification actually occurs in a hot water cylinder. If you have ever looked at an electric element when it is heating a kettle of water, you will notice that the circulation flow of water coming off the heating element is turbulent, causing all the water above the heating element to mix and so no stratification occurs as the water is all mixed. Likewise, when a solar or boiler heating system is heating a hot water cylinder through a heat exchanger, the flow of heat off the coil is thought to be turbulent and so mixing of the hot water above the heat exchanger is thought to minimize stratification. However, when a hot tap or point of use is opened, a smooth flow of cooler water enters the bottom of the hot water cylinder and this cooler water creates stratification within the cylinder. This science is not fully understood and the rate of flow of either heat or cold water into the cylinder will affect the performance of the unit and so how much and when the cylinder stratifies. However, in most cases it is thought that opening a tap creates stratification and heating the cylinder tends to de-stratify the unit.

Types of solar vessels

There are several different ways of storing hot water for solar domestic hot water systems. They can be vented, with one or two low pressure cylinders; unvented and fed directly with cold water from the mains; a thermal store where solar heated water is stored and the cold mains water is heated via a heat exchanger; and there are ways of using solar energy with a combi boiler. There are also instantaneous heaters (combi boilers) with solar input, but with no or limited back up storage. Each of these is looked at in more detail in this chapter. There are also other configurations that are possible too.

Different types of solar vessels

ACTIVITY 14

In your own words, explain why solar instantaneous water heating wouldn't work and why solar storage is a necessity in a domestic solar water heating system. Also explain why the system needs a back up source of heating.

VENTED AND UNVENTED STORAGE OPTIONS

Vented and unvented storage vessels

Both twin-coil, and single-coil storage vessels can be manufactured for use in a vented or unvented solar hot water system. The difference between these two systems is water pressure.

Unvented systems can deliver hot water at mains pressure to all outlets, which is not the case for **vented systems** which have a cold water tank in the attic. However, **unvented systems** have much stricter regulations controlling their construction and installation.

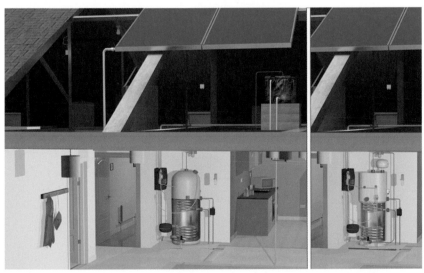

Difference between twin-coil and single-coil storage vessels is water pressure

Twin-coil storage vessels

The **twin-coil storage vessel** has one coil heated from the solar source, and another heated by the back up heating system, to be used when the solar contribution is low. The solar coil is lower and

Vented storage systems These have a cold water tank in the attic

Unvented storage systems Unvented storage vessels do not have a cold water tank in the attic and can deliver hot water at mains pressure to all outlets. Unvented systems are subject to greater regulations than vented systems. Twin-coil and single-coil storage vessels can be manufactured for vented or unvented solar hot water systems

Twin-coil storage vessel The twin-coil storage vessel has one coil heated from the solar source, and another heated by the back up heating system, to be used when the solar contribution is low therefore heated water will rise in the vessel to be available when hot water is required. The greater the separation between the solar and back up heating coils the more efficient the solar contribution will be. A control system for the back up system is needed to prevent use when the solar contribution is sufficient. Solar sensor pockets should be located by the solar heat exchanger.

- Single storage vessel with separate heat exchange coils for solar and back up sources
- Less floor space required, subject to height, so it can occupy the same footprint as the original hot water cylinder
- Solar energy available without delay, as hot water gravitates to top of the cylinder, with short residence times which reduces potential heat loss
- Back up immersion heaters or boilers must have temperature and time control to prevent contributions during useful periods of solar heating
- Can be vented, unvented or thermal store
- Solar sensor pockets located adjacent to solar heat exchanger
- Good insulation, the thicker the better

The vessel's size must sufficiently meet the collector's needs

Twin-coil storage vessel

Size of vessel

The size of the vessel must be sufficient to meet the needs of the collector, plus meet the needs of the household without any solar contribution. As a result twin-coil cylinders sizes have total volumes larger than that required to meet normal total daily requirements. Typical cylinder volumes for solar domestic hot water systems with flat plate collectors are shown in the table:

Solar panel area (flat plate collector)	Solar panel area (evacuated tube collector)	twin-coil cylinder capacity
3m^2	2.4m^2	140–160 litres
4m^2	3.2m^2	200–210 litres
6m^2	4.8m^2	280–300 litres

Back up volume (V$_B$)

The volume heated by the back up heater must be sufficient to meet the needs of the household when the solar contribution is low or non-existent.

Solar volume (V$_s$)

The volume heated by the solar collector must be sufficient to match the surface area of the collector. Approximately 1m^2 of solar flat plate requires a volume of 50 litres of water to heat and approximately 1m^2 of evacuated tube requires a water volume of 62.5 litres.

For example, a 3m^2 flat plate would require either approximately a 150 litre preheat cylinder or 150 litre twin-coil cylinder. The twin-coil provides 150 litres of solar water to heat when the back up heating system is switched off by the central heating programmer.

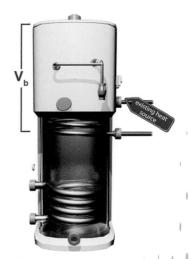

Back up volume

Solar volume

> **Solar Volume (V$_s$)**
> The volume heated by the solar system must match the surface area of the collector. Approximately one square metre of flat plate collector requires 50 litres of water

---

E-LEARNING

Use the e-learning programme to see more information about twin-coil storage vessels.

Preheat cylinder (single coil solar) storage vessels

A separate hot water cylinder heated only by solar energy via a single coil can provide improved solar efficiency because the tank is fed by the cold water supply and no additional heating to this volume is provided by a back up source. The existing hot water cylinder is fed by water from this preheat vessel. The preheat vessel can be positioned in a convenient location depending on available space. There is greater hot water residence time because solar preheated water has to pass through the existing hot water cylinder before being delivered to the taps and this can cause greater heat loss which might offset the solar gains.

Single coil solar storage vessels

> Most commentators believe that the heat losses from having two cylinders outweigh the efficiency gains from having two separate cylinders. The author believes that careful operation of a twin-coil solar system is in most cases the most efficient system. However, this means that the householders must understand and use best practice to maximize their solar gain.

Preheat Volume (V$_s$)
The volume heated by solar energy should be at least 80 per cent of the household needs and the pattern of usage affects its efficiency

- Two storage vessels; one preheat with a dedicated solar volume and single coil from the collector, and the existing heat source cylinder with a single exchange coil from the back up system
- Cylinders can be close together or separated depending on available space and potential heat loss
- Greater residence time, so high levels of insulation needed to reduce increased heat loss.
- Can be vented, unvented or thermal store
- Solar sensor pockets located adjacent to solar coil
- Good insulation required

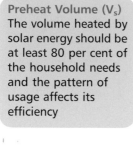

Back up volume

Preheat solar volume

FUNCTIONAL SKILLS

Size of existing vessel, back up volume (Vb)
The existing heat source vessel will have a volume, Vb which is probably correct for the size of household. Rough estimates of 40 to 45 litres per person per day can be used.

Preheat solar volume (V$_S$)

The volume of the solar preheat cylinder should not be less than 80 per cent of the average daily hot water usage. The efficiency of the system is affected by the pattern of usage, if most hot water is used after sunset and before mid-morning a smaller preheat cylinder can reduce efficiency significantly. A table showing typical vessel sizes relating to solar collector is shown:

Solar panel area (flat plate collector)	Solar panel area (evacuated tube collector)	Preheat cylinder capacity
3m^2	2.4m^2	at least 135 litres, ideally 150 litres
4m^2	3.2m^2	at least 170 litres, ideally 200 litres
6m^2	4.8m^2	at least 250 litres, ideally 300 litres

ACTIVITY 15

In the box above, it was said that 'The author believes that careful operation of a twin-coil solar system is the most efficient system. However, this means that the householders must understand and use best practice to maximize their solar gain'.

How would you advise the occupants to maximize their solar gain with a twin-coil cylinder?

Unvented solar hot water cylinders

Unvented solar cylinders are gaining popularity because they deliver mains pressure hot water to all outlets. The pressure is balanced so showers do not need supplementary pumps and there is no cold water storage needed in the roof. However, unvented hot water storage has stricter regulations regarding construction and installation.

Unvented solar cylinders can deliver mains pressure hot water to all outlets

HEALTH AND SAFETY

The three levels of safety on unvented cylinders
The safety requirement is that there must be three levels of safety for each heat energy input.

- Level one is a control thermostat for the energy source with automatic thermostat cut-out at 60°C–80°C

- Level two is the ability to manually reset an energy cut-out set between 80°C and 85°C

- Level three is a Temperature and Pressure Relief Valve (called a TPRV) normally set at 90°C–95°C and 8–10 bar pressure

  Level 2 solar control is discussed in more detail in the next topic.

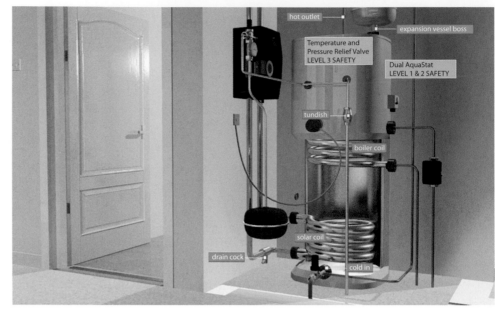

There must be three levels of safety for each heat energy input

UK AND INTERNATIONAL STANDARDS

Building regulations and unvented cylinders

As well as the safety levels all unvented solar hot water systems with a storage cylinder exceeding 15 litres require a competent installer, notifications made to the building controller and water undertaker, and there must be sufficient mains water expansion capability.

- Requirements laid out in BS7 206 and Approved document G of the building regulations

- Competent Installer (Building Reg. G3)

- Notification required to building control officer and water undertaker

- Sufficient mains water expansion capability

ACTIVITY 16

The safety standards in the UK and Ireland for unvented hot water cylinders are particularly onerous. The particular regulations for England and Wales are laid out in the Domestic Building Services Compliance Guide, which can be downloaded from the Internet. Most European countries have a lower level of installation standard. The consequences of an accident with an unvented cylinder are potentially very serious and the more risk averse policies in play across the UK encourage these higher unvented safety standards. There are not a significant number of incidents associated with unvented cylinders reported across Europe. Please set out what is the most significant consequence of an incident with an unvented cylinder and the advantages and disadvantages of the higher safety levels.

Level 2 control on unvented solar hot water cylinders (a British rather than European requirement)

Due to the much higher temperatures, Level 2 control on unvented solar cylinders has different recommendations for the solar coil. There are two options:

Option 1 can be used if the solar collector is located completely above the hot water cylinder. Suitable high temperature and glycol resistant check valves should be fitted to the solar circuit. The electrical supply to the solar system pump should be run via the non-self-resetting thermal cut-out on the dual aquastat so that when the thermal cut-out is activated, the pumped circulation stops.

Option 2 has to be used if the collector is level with or below the hot water cylinder. A spring-closing steam grade two-way valve should be installed on the solar circuit flow within 1m of the cylinder or a spring-closing solar grade two-way valve should be installed on the solar circuit return within 1m of the cylinder.

On no account should a central heating grade two-way valve be used on a solar circuit as these valves are only designed for central heating circuits.

Level 2 control on unvented solar hot water cylinders

E-LEARNING

Use the e-learning programme to see a demonstration of the Solar Grade.

Solar primary circuit
Solar primary circuits connect the energy generation (the solar collector) with energy storage (hot water cylinder) and provide a means of transferring heat from one to the other

ACTIVITY 17

What might happen if a central heating two-way valve was fitted into a **solar primary circuit** 1m from the cylinder on the flow circuit?

THERMAL STORES

Open vented thermal store

A thermal store provides mains pressure hot water to all taps but is not typically under the unvented cylinder regulations because the volume of stored hot water under pressure is less than 15 litres. A thermal store has a large volume of hot water, typically between 50°C and 80°C and this volume of hot water is normally open vented. Mains pressure cold water is passed through a large and very efficient heat

Open vented thermal store

exchanger and exits through a thermostatic mixing valve before passing through to the hot taps. The insulation needs to be very efficient.

UK AND INTERNATIONAL STANDARDS

Please note that a sealed thermal store (i.e. not open vented) is under Part G3 unvented cylinder regulations as the sealed water volume is greater than 15 litres

- Low pressure cylinder
- Mains pressure hot water within a heat exchange coil
- Solar sensor pockets
- Good insulation

Solar coil

Non-solar thermal stores typically operate between 50°C–80°C. So if a thermal store is used as part of a solar domestic hot water system the solar collectors must get hotter than the coolest part of a thermal store before any contribution can be made. Therefore, it is important that a solar thermal store is designed with a significant dedicated solar volume.

Solar coil

Overflow tank

This thermal store is vented but other arrangements are possible. If the thermal store itself is unvented all the associated regulations apply.

- A solar thermal store can be used as either a solar twin-coil or solar preheat two cylinder layout

Considerations when selecting a thermal store

A thermal store has some useful advantages additional to those already mentioned. As the volume of hot water held in the heat exchanger is relatively small and used frequently and at a high temperature, risk of Legionella is much lower. The thermal store can be used to supply underfloor space heating as well as hot water.

Disadvantages are the complexity of some designs and also questions about solar heat transfer from the solar primary volume. Good design should overcome these disadvantages.

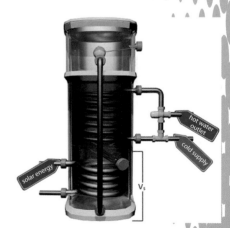

Overflow tank

Risk of Legionella is much lower

Additional considerations when selecting a thermal store:

- Good Legionella control
- Can also be configured to provide small solar contribution to space heating demand
- Often complex designs
- Auxiliary heating required to operate most of the year

ACTIVITY 18

The primary circuit is the circuit providing the heat and the secondary circuit is the circuit absorbing the heat. We know from our study of heat that it always flows from hot to cold. We also know that the primary circuit needs to be around 5°C warmer than the secondary circuit and the higher the temperature difference between the primary circuit and the secondary circuit, the faster the rate of exchange of heat. Using these principles, explain why the primary to secondary heat exchange coil will typically be much greater in a thermal store as compared to a vented or unvented cylinder.

Issues for combi boilers and solar

A range of issues can arise when an existing combi boiler is used within a solar domestic hot water system without extra adaptations. The combi boiler provides mains pressure hot water and space heating via two separate circuits in the boiler. Without taking precautions a variety of problems can arise, including damaging the boiler hot water circuit components, allowing a build up of Legionella and scale deposits.

Issues that can arise when supplying solar preheated water to an existing combi boiler:

Flow rate The rate at which the heat transfer fluid travels around the circuit. The flow rate needs to be set according to the manufacturer's instructions and the pump needs to be set at the lowest setting that can achieve that rate

- Solar preheat temperatures can damage boiler components
- Boiler lockout from over reaction of gas valve
- Excessive water temperatures from overshoot on gas valve control
- Poor potable water sterilization at high **flow rates**

- More scale deposits in boiler heat exchanger
- Invalidation of boiler warranty
- Potential for solar heat loss to cold boiler heat exchanger

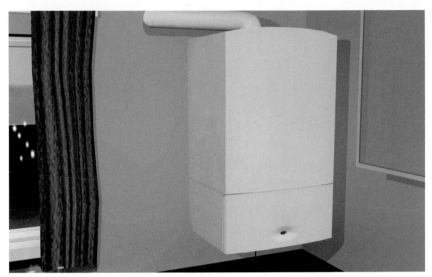

Problems can arise when using a combi boiler with a solar system

Options for an existing combi boiler

An existing combi boiler can be used within a solar domestic hot water system provided the issues we just looked at are addressed. The warranty needs to be checked and, provided that is OK, a solar heated storage vessel can be installed and various changes need to be made to the hot water circuit.

Either:

- Check combi boiler warranty conditions and if allowed by the manufacturer, fit a solar preheat cylinder

Or

- Install twin-coil mains pressure cylinder (i.e. unvented or thermal store)
- Combi boiler to heat top coil of cylinder by changing combi boiler space heating circuit to a 'Y' or 'S' plan
- Reconfigure the distribution pipes from the combi boiler domestic hot water circuit to supply hot water to one low demand tap. (to maintain combi domestic hot water circuit)

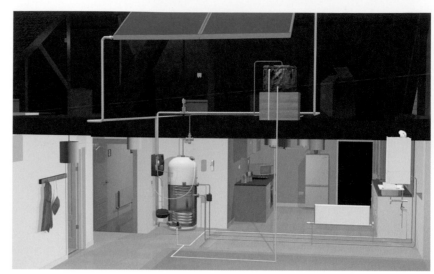

You can use a combi boiler with a solar system providing issues are addressed

ACTIVITY 19

Some manufacturers have solar ready combi boilers available on the market and others have developed designs using blending valves and combi boilers to make sure that water above the boiler trip-out temperature never reaches the combi boiler. This workbook is keen to promote innovation in solar design and also to encourage robust solar systems that are safe for end users and also for installers to fit and commission. Thinking about Legionella issues, what are the particular issues that could occur with a preheat solar store fitted to a combi boiler?

CALCULATING SIZES

FUNCTIONAL SKILLS

Solar storage vessel volumes

Working out the right size of storage vessels and collector surface areas is vital for the system to work efficiently. The important volumes to measure vary slightly according to the system design. This is a twin-coil storage vessel, the solar volume, V_s is the lower part of the cylinder, and the back up volume, V_b is the top part. Together they give the total volume of stored hot water, V_t. The daily hot water demand, V_d, depends on the size of house or household, and is commonly estimated at approximately 40 to 45 litres per person per day. All are measured in litres.

V_s = dedicated volume of water that can be heated by solar energy
V_b = volume of water that can be heated by a boiler or other heat source
V_t = total volume of stored hot water
V_d = the daily hot water demand

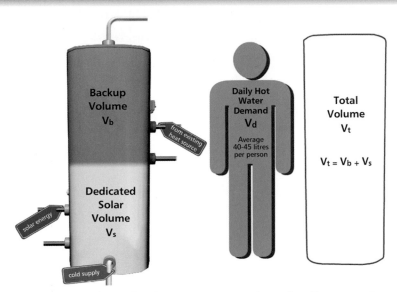

Working out the right size storage vessels and collector surface area is vital

The system shown in the diagram is a preheat cylinder system. The dedicated solar volume and back up volume are now completely separated and the total volume, V_t, is the sum of both storage vessels. The daily hot water demand remains the same

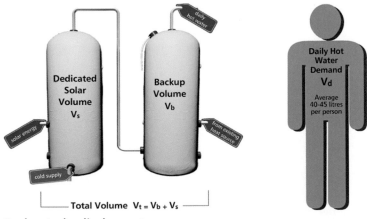

Preheated cylinder system

E-LEARNING

Use the e-learning programme to learn more about the Solar Storage Vessel Volumes.

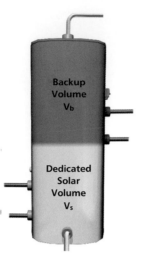

The volume can be calculated from the household and number of occupants

Calculating storage vessel volumes with a twin-coil cylinder

The volume required for the storage vessels can be calculated from the household and the number of people living there. Peak flow when most hot water is being used at any one time is used to calculate the minimum size of the back up volume; the solar volume can be calculated from that – or from the size of the collector.

FUNCTIONAL SKILLS

- The collector size should be balanced with the energy needs of the storage vessel and user demand.

- Peak flow (maximum hot water draw-off) should be met by the back up heat supply (V_b)

- V_s calculated in one of two ways:

  - allow at least 25 litres of V_s per square metre of solar collector

  - OR 80 per cent of the daily demand (80 per cent V_d)

The worked example uses the volumes we have discussed. Check you can follow the method. Additionally you may wish to look up the supporting material from the regulations.

FUNCTIONAL SKILLS

An example:
Assumptions:

    three persons in a house

    one square metre of solar collector per person (= $3m^2$)

    one standard bath and one shower in the house

    Peak demand is about 100 litres

V_d = three people × 45 litres per day (= 135 litres)
V_b = 100 litres (peak demand)
V_s = V_d × 80 per cent (= 108 litres) OR V_s = $3m^2$ collector × 25 litres
    (= 75 litres)

Total cylinder volume V_t = 100(V_b) + 75(V_s) OR 100(V_b) + 108(V_s)
Total cylinder volume should be between 175 and 208 litres

Allow 25 litres of Vs per square metre of solar collector

Water oulets

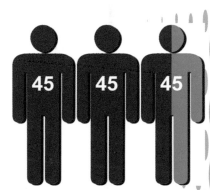

VS calculated as 80% of daily demand

Calculating storage vessel volumes for solar preheat systems

In the previous twin-coil example, there is 100 litres of dedicated solar storage when the back up heating is on and 175 to 208 litres of solar storage when the back up heating is switched off. With a preheat solar cylinder, the dedicated storage volume should typically be twice the dedicated solar volume of a twin-coil cylinder

- The collector size should be balanced with the energy needs of the storage vessel and user demand

- Peak flow (maximum hot water draw-off) should be met by the back up heat supply (V_b)

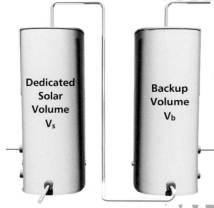

Balance the energy needs and user demand

- V_s calculated in one of two ways:
    - allow at least 50 litres of V_s per square metre of solar collector
    - OR 80 per cent of the daily demand (80 per cent V_d)

The worked example uses the volumes we have discussed. Check you can follow the method. Additionally you may wish to look up the supporting material from the regulations.

FUNCTIONAL SKILLS

An example:

Assumptions:

   three persons in a house

   one square metre of solar collector per person (= 3m²)

   one standard bath and one shower in the house

   Peak demand is about 100 litres

V_d = three people × 45 litres per day (= 135 litres)

V_b = 100 litres (peak demand)

$V_s = V_d$ × 80 per cent (= 108 litres) OR $V_s = 3m^2$ collector × 50 litres (= 150 litres)

Total cylinder volume $V_t = 100(V_b) + 150(V_s)$ OR $100(V_b) + 108(V_s)$

The back up cylinder volume should be over 100 litres and the solar cylinder volume would ideally be 150 litres with a minimum of 108 litres.

Remember that if you are calculating a system with an evacuated tube collector they are more efficient and you would need to use a different value for the area used per person. 1m² of flat plate collector is equivalent to 0.8m² of evacuated tube collector, so our example above would use:

   0.8 square metre of solar collector per person (making the collector area 2.4m²)

Therefore $V_s = V_d$ × 80 per cent (= 108 litres) OR $V_s = 2.4m^2$ collector × 62.5 litres (= 150 litres) thus the solar cylinder volume would ideally be between a minimum of 108 litres and 150 litres

Remember that if you are calculating a system with an evacuated tube collector they are more efficient and you would need to use a different value for the area used per person.

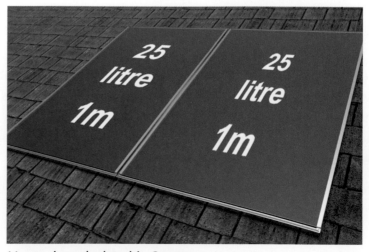

Vs can be calculated in 2 ways

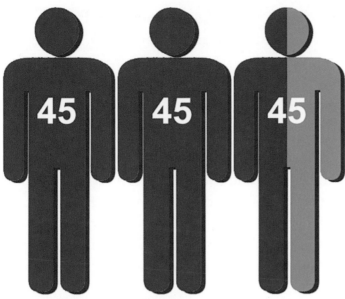

VS calculated as 80% of daily demand

Water oultets

ACTIVITY 20

Calculate V_s, V_b, V_d and V_t for a twin-coil and two cylinder solar system for a detached house with four bedrooms and two bathrooms. Assume the occupancy is number of bedrooms plus one and please note that BS 6700 indicates that a two bathroom house requires at least 130 litres of storage.

Solar heat exchange surface area

- Flow rates of **less than 0.5 litres per minute** per square metre of solar collector need a coil surface area of **0.1m²** per collector square metre

- Flow rates equal to or **more than 0.5 litres per minute** per square mete of solar collector need a minimum coil surface area of **0.2m²** per collector square metre

This advice is taken from the English and Welsh Domestic Building Services Compliance Guide (DBSCG). It is a recommendation for heat exchanger sizing. In reality heat exchanger sizing is a complex engineering issue because of the significant number of variables that apply during this design process. However, this DBSCG acts as a useful guide for demonstrating cylinder heat exchanger area sizing.

FUNCTIONAL SKILLS

An example:
Collector surface of 3m² with a manufacturer's stated flow of 0.8 litres/min the MINIMUM SIZE of the solar heat exchanger must be:

0.8 litres/min > 0.5 litres/min.
Therefore use 0.2m² coil area per m² collector
0.2 × 3 = 0.6m² minimum coil area

ACTIVITY 21

Specify a coil size for a 4m² flat plate collector with a flow rate of two litres/min.

HAZARDS

What are the hazards?

Working with cold, warm and hot water, which may or may not be moving, can bring hazards of its own. We will consider three in particular:

- Legionella
- Limescale
- Scalding

Although each is different they can all be dangerous in their own way. Each hazard is different but each can be very dangerous and needs to be minimized or removed from the solar system and monitored during maintenance.

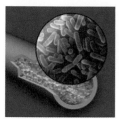
Legionella

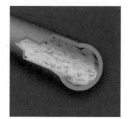
Limescale

Scalding

Legionella

Legionella is a bacteria and particularly likes water with nutrients at about 38°C to 40°C, and it can develop in stored water between 20°C and 50°C. It is found in jacuzzi baths and shower heads, and can also live in domestic hot water systems where favourable conditions exist. It causes Legionnaires' disease, a severe form of pneumonia which is a particular risk to those with low immunity. The most effective way to get rid of it is to have the water temperature above 60°C, when the bacteria are killed within several minutes. Other ways to kill them is to ensure there is no stagnation of the hot water (regular use and no dead legs on pipework), use copper pipes (which are toxic to Legionella) and there is regular maintenance of the system, especially with chlorinated water.

Legionella

 HEALTH AND SAFETY

Please note that:
V_b must be heated to 60°C for one hour per day and the hot water circuit should be checked with a risk assessment for Legionella and other safety risks.

ACTIVITY 22

When discussing Legionella bacteria pasteurization, 90 per cent of this bacteria is said to be killed after water storage at a temperature of 60°C for 32 minute. How much of the Legionella bacteria will be killed if the water is stored at 60°C for 64 minute and what percentage of Legionella bacteria will still be

(Continued on next page)

in the sample as compared to the sample before it was heat treated for 64 minutes? Do you think this percentage of bacteria that was left over could still kill someone?

Limescale

Limescale is formed from a chemical dissolved in the water, particularly in chalk and limestone areas of the country. You can check if you are in an area likely to suffer from limescale by looking in the kettle. If there are deposits on the heating element you have a problem. Kits are available to test the problem, or a local water authority report. Limescale can significantly reduce heat transfer and increase fuel costs by around six per cent per mm of scale deposit. Various methods of solving the problem can be found, and the best for the system can be identified after the water hardness has been assessed.

- Protection of new and existing services in areas of high risk
- Some makes of water conditioner
- Water softener
- Control hot water temperature; no more than 65°C and preferably 60°C
- Chemical cleaning to remove scale
- In a soft water area, as long as scalding considerations are taken into account, the solar controller is often set to a maximum temperature of 75°C or 80°C to gain more solar energy

Limescale

ACTIVITY 23

Boiler temperature is normally set with a cylinder stat and Tmax in the solar system is normally set up in the solar controller (also called a DTC). What is the difference between these two stats/sensors systems and how do you think they might work/function?

Scalding

HEALTH AND SAFETY

Solar hot water systems are capable of producing water at very high temperatures, even above normal boiling point in pressurized systems. To prevent users being scalded by the water, thermal mixing valves – often called TMVs or blending valves – could be used. This is particularly important where there are people categorized at risk, such as the elderly, children and the very young.

There are three strategies for managing scalding risk:

1. Fit a TMV set at 43°C (or other local required temperature) within 1m of all the points of use

2. Fit a TMV close to the cylinder that distributes hot water at between 55°C and 60°C

3. Or limit the maximum temperature of the solar controller to 65°C and preferably to 60°C

All the equipment used in solar systems must be capable of withstanding high temperatures without distorting or deteriorating.

- Most available collectors are capable of reaching temperatures of between 150°C and 300°C

- Use of thermal mixing valves prevents the hot water delivered by the solar system becoming too hot – particularly where there are persons categorized at risk

- Current recommendations to store water at 60°C makes use of thermal mixing valves (also called anti scald or blending valve) more appropriate

- Thermal mixing valves are susceptible to dirt and debris and therefore water must be strained/filtered

- Thermal mixing valves can only tolerate a small imbalance of pressure between hot and cold inlets, so use of a pressure reducing valve may be necessary

All equipment used in solar systems must withstand high temperatures

HEALTH AND SAFETY

Scald injuries are especially common in children less than five years old, adults over 65 and anyone with poor health. Each year across the USA (total population 270 million), about 112 000 people are treated in hospital emergency rooms with scald burns and about six per cent of them are kept in hospital. Many of these scaldings are a result of household water being delivered at temperatures above 49°C. About 80 per cent of hot tap water burns are among young children, the elderly and the physically impaired.

Minimizing the hazards

If the system is fitted in a hard water area, it is best practice to limit Tmax (on the solar controller) to 60°C to obtain a balance between Legionella, limescale and scalding requirements. This is because 60°C kills Legionella, limits limescale production and adds no greater risk of scalding. However, where limescale risk is much less significant, as in a soft water area, if so desired to obtain the extra solar gain, Tmax can be increased up to 80°C. If this is done, the end user must be protected from scalding by installing blending valves which can be placed in the primary circuit to ensure the water in the pipework remains between 55°C to 60°C to minimize the Legionella risk.

Please also note that storing and distributing the hot water between 55°C and 65°C is definitely in the scalding water temperature region and so if there is a risk of scalding at any point of use, especially a bath or similar, always fit a TMV set to a much cooler temperature at the relevant points of use.

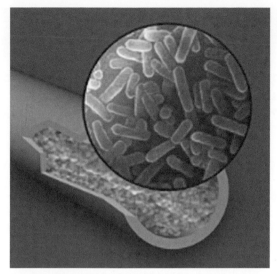

Legionella

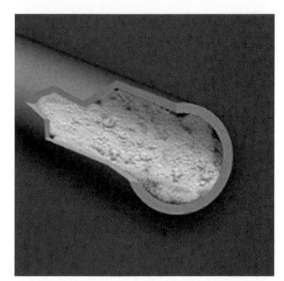

Limescale

Scalding

CHECK YOUR KNOWLEDGE

1. **Which of the following statements apply to unvented systems and which apply to a vented solar hot water system? Place the correct statements in the relevant table.**

☐ a. Can deliver hot water at mains pressure to all outlets

☐ b. Have a cold water tank in the attic

☐ c. Have strict regulations controlling their construction and installation

☐ d. Do not need supplementary pumps

Unvented	Vented

2. **Which of the following regulations apply to unvented solar hot water systems?**

☐ a. All energy sources to have 3 levels of safety

☐ b. All water temperatures to be maintained at 65°C minimum

☐ c. All outlets/inlets in storage tank to be fitted with pressure relief valves

☐ d. All outlets/inlets in storage tank to be fitted with thermal mixing valves (blending valves)

Chapter 5

SOLAR THERMAL PRIMARY CIRCUIT DESIGNS

LEARNING OBJECTIVES

By the end of this chapter you will be able to:

- Explain the purpose of the solar primary circuit

- Identify the factors that need to be addressed by the primary circuit

- Identify the different types of primary circuit

- Describe the importance of flow rates through a primary circuit

PRIMARY CIRCUIT DESIGNS

Primary circuit designs in solar domestic hot water systems

This diagram illustrates the position of the solar primary circuit and its function to connect the energy captured by the solar collector with energy stored in the hot water cylinder. Essentially it provides a means of transferring heat from one to the other with minimal loss in between the two points.

Solar primary circuits connect the energy generation (the solar collector) with energy storage (hot water cylinder) and provide a means of transferring heat from one to the other.

Position of the solar primary circuit

Types of primary circuit designs

There are a number of primary circuit designs that are possible; the first distinction is whether the primary circuit is direct or indirect.

Is the primary circuit direct or indirect?

Direct systems have the water from the domestic hot water store circulated through the collector then stored until used. The circulating fluid and domestic hot water are the same.

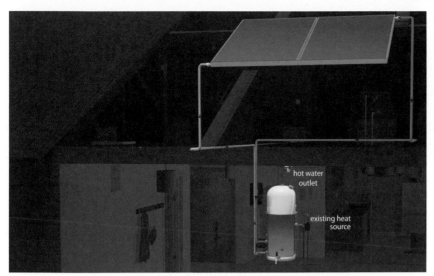

Indirect systems - circulating fluid and domestic hot water never mix

Indirect systems are very different because the circulating fluid and domestic hot water never mix. The primary circuit passes fluid through the collector and then it transfers the collected heat to the hot water store through a heat exchanger (e.g. coil in storage vessel).

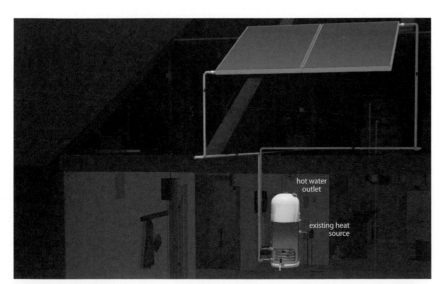

Indirect system

E-LEARNING

Use the e-learning programme to see a demonstration of direct and indirect systems.

ACTIVITY 24

Direct heating systems connected to gas or oil boilers went out of use over 50 years ago across northern Europe because of the hygiene advantages of separating the dirty water in the central heating circuit from the potable drinking water in the secondary hot water circuit via a heat exchanger. How do you think solar direct water circuits which are installed today overcome this issue of keeping the water potable in a direct heating system?

The main type of primary system in use in the UK and across northern Europe is the indirect primary circuit design, where the fluid circulating in the collector transfers its heat to the domestic hot water via a heat exchange coil in the hot water cylinder. There are two types of indirect systems in use in the UK which operate in rather different ways: fully filled pressurized primary circuits and drainback primary circuits. We shall look at each of them in turn.

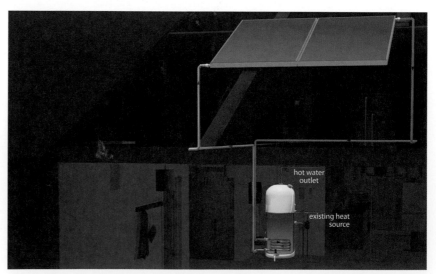

hot water outlet

existing heat source

The indirect primary circuit design is the most popular primary system

Primary circuit designs

Primary circuits are subject to the same hazards as the rest of the system, and we have already come across a similar list earlier in the workbook. Here we need to consider six issues in particular: extremes of heat and cold, limescale, bacterial growth, expansion of fluids and limiting energy losses.

All types of solar primary circuit must be capable of:

- Withstanding frost
- Managing high temperatures and steam
- Limiting the effects of limescale
- Controlling bacterial growth (including Legionella)
- Managing fluid expansion without user intervention
- Limiting energy losses

Primary circuits are subject to the same hazards e.g. limescale

ACTIVITY 25

Without reading on through the book, think about the various issues above and how you might design and install a solar primary circuit to manage the various points.

FULLY FILLED PRESSURIZED SOLAR PRIMARY CIRCUITS

A fully filled pressurized solar primary circuit is shown. The heated fluid leaves the solar collector and travels to the solar heating transfer coil in the storage vessel. Heat transfer takes place and the cooler fluid is then pumped back towards the collector. The primary circuit has a variety of components:

- Collector sensor
- Air vent with mechanical isolator
- 230v electrical supply
- **Differential Temperature Controller** (DTC)
- Flow line check valve
- Return line check valve
- Solar sensor
- Boiler circuit heat exchanger
- Back up sensor
- Drain cock
- Two pump isolating valves
- Pump
- **Flow regulator**
- Pressure gauge
- Fill valve
- Safety relief valve
- Expansion vessel and discharge container

Differential Temperature Controller This controller operates by monitoring the three key temperatures via the sensors and turning the pump on or off as appropriate. It requires temperatures to be set to activate and deactivate the pump. These key temperatures are 'T on' (also called ∆T on), 'T off' (also called ∆T off) and 'T max'

Flow regulator Controls the flow in the primary circuit and there is an optimum rate of flow for heat transfer fluid passing through the solar collector

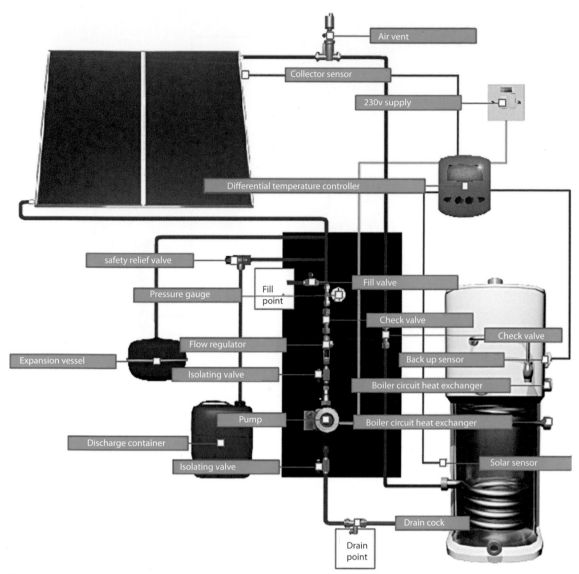

Illustration of a fully filled pressurized solar primary circuit

Pipework

Pipework for this type of indirect circuit can be placed where it is most convenient as no minimum distance or angles are required. It is best to keep the runs as short as possible to maximize the efficiency of the solar circuit.

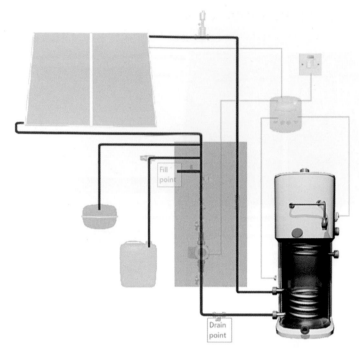

Illustration of pipework

Fluid

The circulating fluid completely occupies the pipework, exchange coil and collector, so there is no air. The fluid is usually a mixture of water and glycol antifreeze and there is not a large volume – so it heats up quickly and starts transferring energy to the storage vessel very quickly.

- **Low fluid content**: the total fluid content of a fully filled pressurized primary circuit could be 6–12 litres, which means less fluid to heat before energy can be added to storage vessel
- **Fully filled with heat transfer fluid**: no air should be present after commissioning. Automatic air vents fitted to the highest point in the system will remove any original air, and as the system is sealed, no further air can enter. The auto air vent should be fitted with a mechanical isolator that is switched off after the system has been commissioned so that any steam generated during stagnation doesn't leave the solar circuit
- **Antifreeze and water mix**: this can be premixed at production or during system installation

Circulation

The circuit operates under pressure, and as the fluid heats up, it expands and increases the pressure. Once maximum hot water temperature has been reached and the solar circuit goes into stagnation, circulation no longer occurs until the collector temperature cools down again.

- **Primary circuit operates under pressure** (variable 1.5 to 6 bar depending on design)

- **No further heat collection at stagnation**: once maximum hot water temperature has been reached, the solar pump is switched off and the solar fluid has converted to steam in the collector, circulation will not be able to resume whilst collector has a 'steam lock'

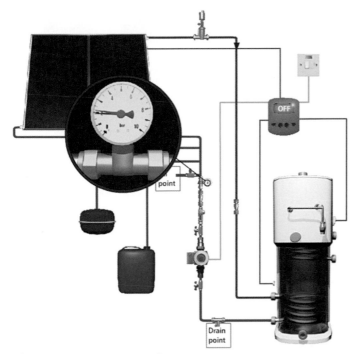

The circuit operates under pressure

E-LEARNING

Use the e-learning programme to see a demonstration of a fully filled pressurized solar primary circuit.

Mediating or solving issues for fully filled primary circuits

The fully filled pressurized solar primary circuit overcomes or reduces the six hazards that need to be addressed so that it can operate safely. Look at each one and, where relevant, look for the features built into the primary circuit to address the issues in the diagram.

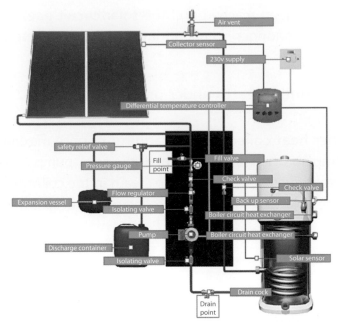

Fully filled pressurized solar primary circuit overcomes the 6 hazards

Frost

The primary circuit is filled with antifreeze, mostly glycol, which will resist freezing. Most modern systems use solar rated non-toxic glycol.

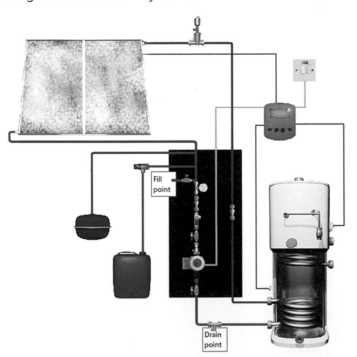

The circuit is filled with antifreeze

High temperature and steam

To manage high temperature and steam the components at risk of damage are carefully positioned in the coolest parts of the circuit, such as the pump and expansion vessel in the return to collector pipe-work. Wherever possible components used are specified to operate reliably at 150°C, e.g. all-metal valves.

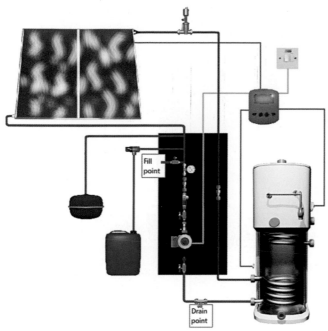

The components at risk of damage are positioned in the coolest parts

Limescale

The effects of limescale are limited because the circuit contains a sealed fluid. With a volume of fluid that does not change, no new source of limescale can enter the system.

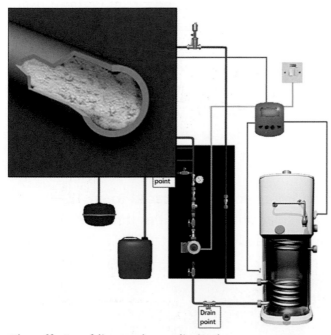

The effects of limescale are limited

Bacteria

There is no direct connection between the fluid in the primary circuit and the hot water that comes out of the taps, therefore there is no issue with bacteria for this type of primary circuit. However, bacteria control methods are still required for the secondary hot water circuit.

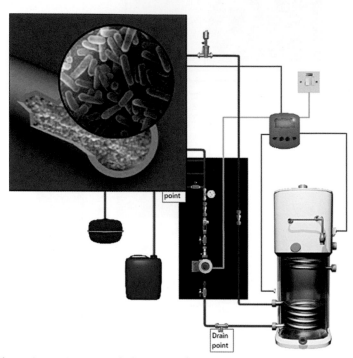

There is no issue with bacteria for this type of primary circuit

Fluid expansion

Managing fluid expansion in a sealed circuit without user intervention requires somewhere for the increased volume to go when the temperature rises. The circuit needs to incorporate an expansion vessel of suitable size to accept all displaced heat transfer fluid from collectors during stagnation without fluid discharge or circuit failure under pressure. We will look at **expansion vessels** separately.

The circuit also includes a safety relief valve as a designed 'safe weak point' in the event of a fault in the expansion vessel or another component.

Note: the maximum volume displaced to an 18 litre expansion vessel in a standard solar primary circuit at three bar fluid pressure is only seven litres approximately.

Expansion vessels
This is a vessel with a membrane inside, which can compensate for volume changes in the primary circuit due to fluid expansion. The vessel is gas filled, and as the fluid expands the gas is compressed, then as the fluid returns to previous volumes the gas pushes the fluid back into the pipework. The drainback system does not operate under pressure and is normally a sealed system

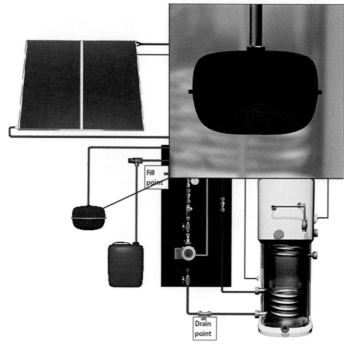

The circuit must incorporate an expansion vessel of a suitable size

Energy loss

Energy loss is reduced by good thermal insulation on all components. The possibility of a **thermosiphon effect** developing – where the hot water is drawn back to the collector – is prevented by using check valves in the circuit, only allowing flow in one direction. Energy loss through ineffective collection of solar energy is corrected by setting the circulation controls correctly.

Limiting energy loss

- Radiant and convection heat loss from pipes and components limited by sufficient insulation

- Radiant, convection and conduction heat loss caused by thermo-siphon within the primary circuit from hot water storage to collector can be limited by correct installation of check valves

- Wasted electrical energy and solar radiation not captured by incorrect circulation control settings

Thermosiphon effect (also spelled thermosyphon) This is where the hot water is drawn back to the collector and is prevented from doing so by using check valves in the circuit, only allowing flow in one direction

E-LEARNING

Use the e-learning programme to see a demonstration of solving issues for fully filled primary circuits.

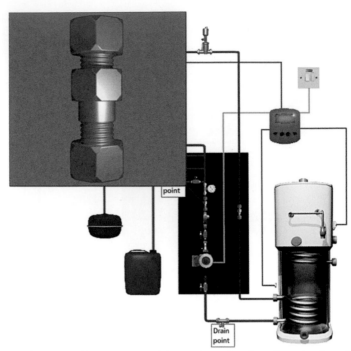

point

Drain
point

Energy loss can be prevented by using check valves in the circuit

Expansion vessels

Expansion vessels come in a variety of shapes and sizes. Essentially they are vessels with a membrane inside which can compensate for volume changes in the primary circuit due to fluid expansion. The vessel is gas filled, and as the fluid expands the gas is compressed, then as the fluid returns to previous volumes the gas pushes the fluid back into the pipework. It is important that the vessel has a maximum working pressure which must not be less than the safety relief valve. The expansion vessel must be big enough to accommodate the total 'steam' volume of the solar primary circuit:

- Gas filled and has a flexible membrane dividing the vessel
- Fluid filled section has a small volume at normal temperatures but as solar heating and stagnation occurs the fluid expands and compresses the gas until there is equal pressure on both sides of the membrane.
- On cooling gas will push fluid back into circulation through the pipework
- Maximum working pressure must not be less than the safety relief valve.
- Size 18–25 litre vessel is sufficient for most domestic installations. The collector supplier or system specifier will confirm the required size.
- Expansion vessel volume = Total steam volume of primary circuit × 2

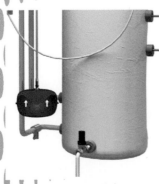

Expansion vessels

ACTIVITY 26

The expansion vessel has to be sized to twice the total steam volume of the solar primary circuit plus a few extra litres as a safety factor. The steam in the solar primary circuit collects in:

- either the total internal volume of the solar collectors plus the internal volume of 4m of the pipework

- or the internal volume of the solar collectors plus the internal volume of the flow pipework to the hot water cylinder.

How would you go about calculating these volumes so you could correctly size the solar expansion vessel?

DRAINBACK SOLAR PRIMARY CIRCUITS

The **drainback solar primary circuit** is shown. This system is becoming more popular in the EU and several suppliers are selling this system in the UK. The main feature of the drainback primary circuit design is that the components are specifically positioned to manage temperature extremes and energy loss. The drainback primary circuit has a variety of components:

- Collector sensor
- 230v electrical supply
- Differential Temperature Controller (DTC)
- Drainback vessel solar sensor
- Boiler circuit heat exchanger
- Back up sensor
- Drain cock
- Two pump isolating valves
- Pump
- Flow regulator
- Pressure gauge
- Safety relief valve
- Discharge container

> **Drainback solar primary circuit** The main feature of the drainback primary circuit design is that the components are specifically positioned to manage temperature extremes and energy loss

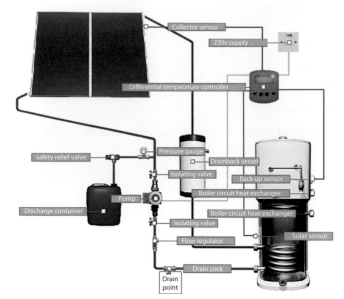

Illustration of the drainback solar primary circuit

Pipework

Pipework for this type of indirect circuit has to be positioned to allow specified pipe angles and vertical pipe distances between the drainback vessel and the collector. This makes it less flexible than a fully pressurized primary circuit. Two requirements that need to be covered are:

- Minimum fall on pipes is normally 30mm per 1m length. The manufacturer will specify the minimum fall angle of pipes between the drainback vessel and the collector

- The maximum height depends on the 'head' of pressure that can be delivered by the solar circulating pump. For example, if a 5m head pump is used, the drainback vessel must be located no more than four vertical metres from the top of the collector

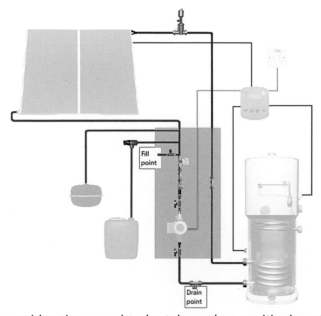

Special considerations need to be taken when positioning pipework

Fluid

The circulating fluid is either water or a water and antifreeze mixture, which is not under pressure. The amount of fluid is important and needs to be accurate for the installation; too much will not be efficient, too little will prevent circulation. These drainback systems need air in the system to function properly. Well designed drainback systems will not increase noise during circulation.

- **Heat transfer fluid** can be water or antifreeze and water mix
- **Fluid content should be accurate**; too much heat transfer fluid than necessary will increase thermal mass to be heated before heat energy can be transferred to the hot water cylinder; too little will reduce or prevent circulation. The heat transfer fluid volume of a drainback primary circuit will often be greater than that of a fully filled pressurized primary circuit. The drainback vessel volume must be greater than the total air volume of the system
- **Drainback systems must have air present** in the solar primary circuit in order to work

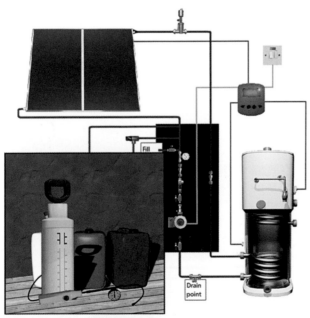

The fluid circulating is water or a mixture of water and antifreeze

Circulation

The drainback system does not operate under pressure and is normally a sealed system. Protection at extremes of temperature occurs by the circulating fluid draining back into the drainback vessel. Even if stagnation has occurred, and hot water is required, the fluid can be pumped to the collector because the fluid will act as a coolant and be able to collect solar energy and heat the domestic hot water system. It is important that only collectors suitable for use in drainback systems are used.

- Protection from temperature extremes: heat transfer fluid in the collector and all pipes outside the thermal envelope are returned to drainback vessel automatically if temperature extremes occur

- Heat collection can occur even at stagnation: if further hot water is required after the collector has stagnated, the pump will begin circulating heat transfer fluid. When fluid first re-enters the stagnated collector, fluid will temporarily convert to steam but will quickly condense as the absorber's low thermal mass is rapidly cooled to below boiling point by the circulating fluid. So at stagnation, further circulation (heat collection) is available

- Collectors must be suitable for use in drainback systems: Collectors suitable for drainback have absorber pipe arrangements that avoid trapping fluid when draining down

E-LEARNING

Use the e-learning programme to learn more about the drainback solar primary circuit.

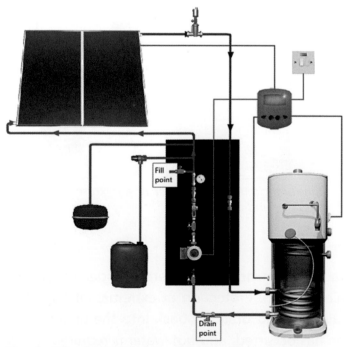

The drainback system does not operate under pressure

ACTIVITY 27

It was stated above that the liquid volume in the drainback vessel has to be greater than the air volume above the drainback vessel. How would you go about calculating the air volume above the drainback vessel?

Mediating or solving issues for drainback primary circuits

HEALTH AND SAFETY

The drainback solar primary circuit overcomes or reduces the six hazards that need to be addressed so that it can operate safely. Look at each one, and where relevant look for the features built into the primary circuit to address the issues in the diagram.

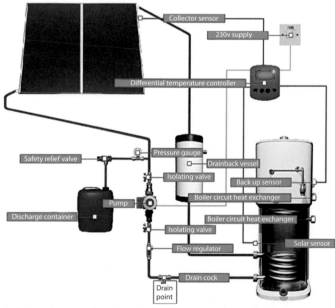

Solving issues for drainback primary circuits

Frost

The sensors on the collector can detect low temperatures near freezing and when this happens the pump is turned off and the circulating fluid will drain back into the drainback vessel. The collector will not therefore be damaged by ice. Some drainback systems also use anti-freeze for double protection.

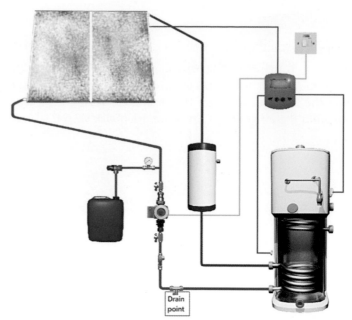

The sensors on the collector are designed to detect low temperatures near freezing

High temperature and steam

To manage high temperature and steam the same control system works to shut down the pump, allowing the fluid to drain back as seen, protecting system components. The pump is turned off and gravity allows the fluid to drain back to a cooler position. Component specification and positioning are similar to fully filled pressurized systems but there are fewer components to protect.

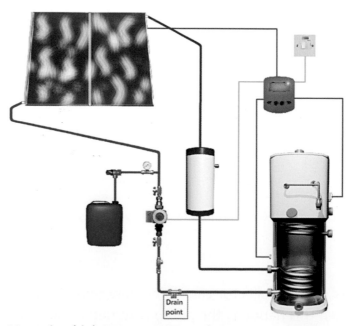

Managing high temperatures and steam

Limescale

The effects of limescale are limited because the circuit contains a dedicated volume of fluid and therefore no new limescale can enter the system.

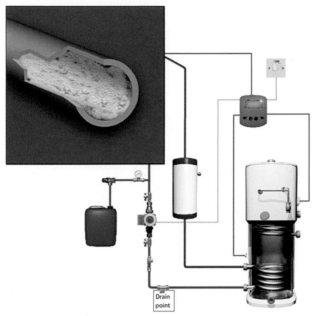

The effects of limescale are limited

Bacteria

There is no direct connection between the fluid in the primary circuit and the hot water that comes out of the taps, therefore there is no issue with bacteria for this type of primary circuit. However, bacteria control methods are still required for the secondary hot water circuit.

When using this type of primary circuit, there's no issue with bacteria

Fluid expansion

The drainback vessel contains air, which moves around the circuit as the circulation starts and stops. As the fluid expands during solar heating, the air in the drainback vessel becomes a compressed air pocket, but because the volume of fluid and size of drainback vessel have been carefully calculated, this pressure should not be excessive. However the circuit also includes a safety relief valve as a 'safe weak point' as a back up.

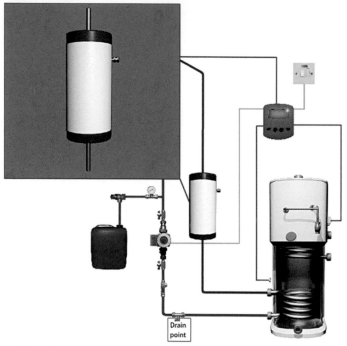

Drainback vessel contains air, moving around the circuit as circulation starts and stops

Energy loss

As well as sufficient insulation, energy loss at night is minimized because energy can only gravitate from storage vessel to drainback vessel as all pipes above this point contain no fluid when the pump is not running.

Limiting energy loss

- Energy loss: at night is minimized as energy can only gravitate from storage vessel to drainback vessel as all pipes above this point contain no fluid when the pump is not running
- Sufficient solar grade insulation
- Correctly set solar controller

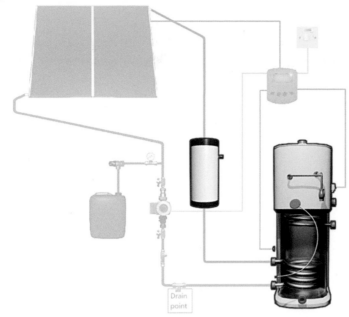

Energy loss at night is minimized as energy gravitates from storage to drainback vessel

E-LEARNING

Use the e-learning programme to see more information about solving issues for drainback primary circuits.

Drainback vessels

Drainback vessels are specified for flow or return pipe connection and come ready insulated. The size required for most domestic solar systems is between 5 and 20 litres. As discussed earlier, the volume of fluid needs to be accurate otherwise the system is inefficient. The vessel is sized so that the water volume of the drainback vessel is greater than the 'air volume' of the pipework above the drainback vessel, plus the internal volume of the collectors.

FUNCTIONAL SKILLS

- Overfilling = more thermal mass in solar primary circuit and slower heat transfer

- Underfilling = shortage of fluid to complete full circulation; no heat transfer

- Typical drainback vessel size = internal volume of pipework above the drainback vessel + internal volume of the collectors + five litres (your drainback supplier will provide exact calculation methodology)

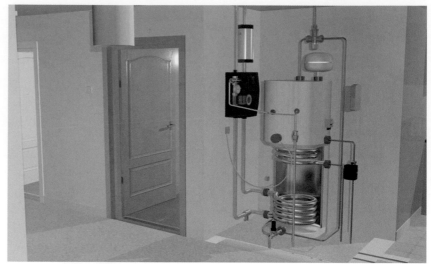

Drainback vessels are specified for flow or return pipe connection

ACTIVITY 28

Discuss the main advantages and disadvantages of solar fully filled and drain-back solar circuits.

SPLIT COLLECTOR PRIMARY CIRCUITS

Not all houses have roofs convenient for maximum solar energy collection; however, it is possible to mount collectors on separate sections if a roof faces east–west. As the Sun rises solar radiation strikes the east-facing collector long before the west-facing collector, in the afternoon and evening.

We know the amount of solar energy collected is maximum if the collector faces south, but it is possible to compensate for reduced available energy by oversizing the collector. Approximately 1m^2 of south-facing collector can be replaced by 1.25m^2 of collector on an east or west-facing roof. There also need to be additional circulation controls.

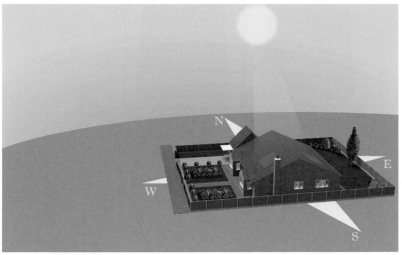

It's possible to mount collectors on separate sections if a roof faces east–west

Split collector solar primary circuit

A **split collector solar primary circuit** is shown. As the Sun rises, solar radiation strikes the east-facing collector long before the west-facing collector, and vice versa in afternoon/evening.

Split collector solar primary circuit Used where collectors need to be mounted on separate sections of roof (east/west-facing roofs)

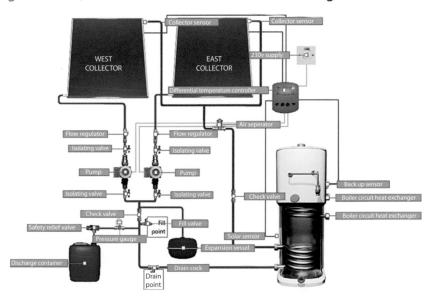

Split collector solar primary circuit

Control

If the system has a single temperature sensing point on the east collector it will begin to circulate transfer fluid to both collectors at the same time and therefore lose heat from the west-facing collector. To ensure optimum energy gain, each collector requires its own temperature sensing and fluid circulation. This in turn requires a more sophisticated controller capable of split temperature sensing (one for each collector) and two power outputs (one for each pump).

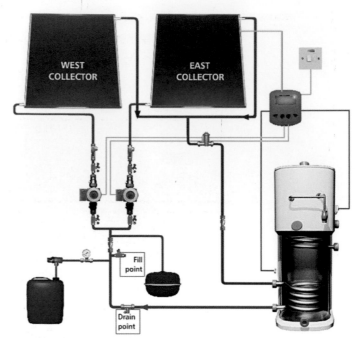

Collectors require their own temperature sensing and fluid circulation

Circulation

A split collector system needs a dedicated fluid circulation for each collector. This can be achieved by either having two pumps or a single pump with a three-port valve or a single pump with two two-port valves. However, if the three-port valve option is chosen this is incompatible with a drainback system as fluid could potentially be trapped in the collector.

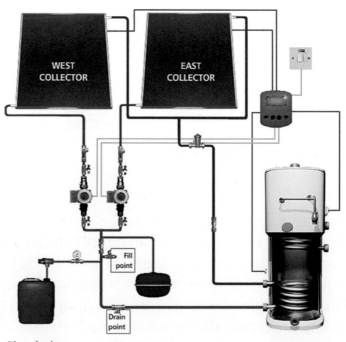

Circulation

Compatibility

The split collector primary circuit can be used with either fully filled pressurized or drainback systems. If a drainback system is used each split collector installation must be able to fully drain in any eventuality (e.g. separate pumps).

Please also note that some manufacturers and suppliers don't recommend a split east–west circuit when faced with an east–west facing roof. Rather they recommend oversizing a single collector on either the west or east-facing roof. Both oversized split and single collector circuits work well and the split circuit will collect energy for a longer time during the day.

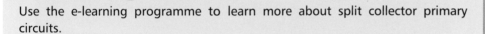

E-LEARNING

Use the e-learning programme to learn more about split collector primary circuits.

ACTIVITY 29

A householder is asking for your advice on comparing two solar systems. There are four occupants in the house. One quote is for an east–west split system with 2.5m^2 of flat plate collector on each pitch of the roof. The other quote is for 4m^2 of evacuated tube collector on the east-facing roof. The split system quote costs fractionally more than the east-facing roof quote. Please explain to the householder the advantages and disadvantages of each system.

FLOW RATES, THEIR IMPORTANCE AND MEASUREMENT

Flow rates

All types of solar primary circuits need a way of controlling the flow of fluid through the collector. There is an optimum flow rate for heat transfer fluid depending on the type of collector and amount of energy in the collector, and the collector manufacturer will always define this optimum rate of flow. It is usually around one litre per minute per m^2 of collector. The rate of flow can be controlled by one of two methods.

The flow of fluid through the collector must be controlled in all systems

Flow control

Flow regulation through the collector can be controlled by either adjusting the speed the pump is being operated, or by using a flow regulator as part of the primary circuit. Most pumps used in solar circuits have three speed settings. A flow regulator is a mechanical restrictor with a measured sight glass and adjustable throttle. Most solar primary circuits use a flow regulator.

Flow control can be achieved by using one of two methods.

- Adjusting the speed settings on the pump:

  - Some solar controllers modulate (adjust the electrical energy to the circulating pump) to control the flow rate. These circuits still need to have correctly commissioned pump and flow regulator settings

- A flow regulator:
  - has a sight glass with flow rate marks in litres per minute and allows visual confirmation of flow
  - has an adjustable throttle

We can either adjust the speed settings or use a flow regulator

ACTIVITY 30

There are 4m^2 of flate plate collector on the roof. The manufacturer suggests a flow rate of 0.9 litres/min/m^2 collector. You have a 5m head solar grade pump (this is identical to a 5m head central heating grade pump except it has a slightly higher temperature specification than the central heating pump). The 5m head pump has three speed settings, one slow 40 watts, two medium 75 watts and three fast 105 watts. What flow rate should the system be set to and which speed setting should you use, and do you think the power consumption at this setting is reasonable or could it be reduced?

CHECK YOUR KNOWLEDGE

1. **Label the numbered parts of the fully filled pressurized primary circuit shown.**

2. **Label the numbered parts of the drainback solar primary circuit shown.**

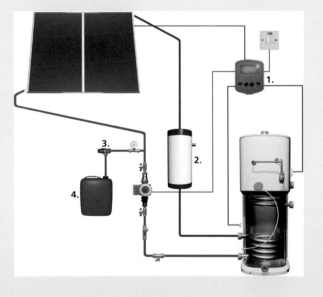

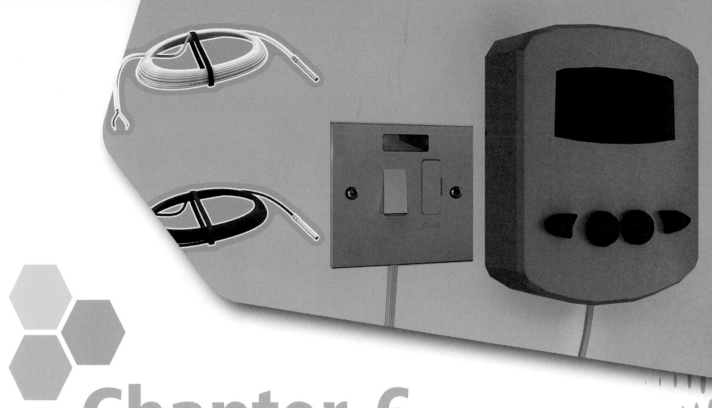

Chapter 6

SOLAR THERMAL PRIMARY CIRCUIT CONTROLS

LEARNING OBJECTIVES

By the end of this chapter you will be able to:

- Describe the purpose of a primary circuit controller

- Describe the types of primary circuit controllers

- Identify inputs required for the controller to operate efficiently

- Describe the operation of the controller

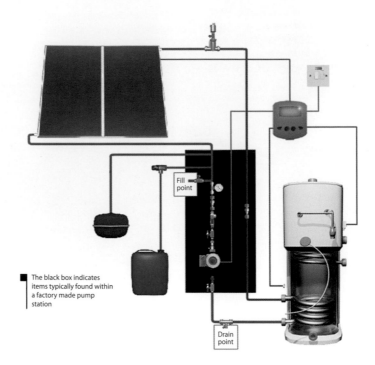

The black box indicates items typically found within a factory made pump station

Solar thermal primary circuit controls

PURPOSE OF A SOLAR SYSTEM CONTROLLER

The purpose of a controller within the solar domestic hot water system is to efficiently control the circulation pump to enable the maximum benefit from the solar energy, and to maintain the user's safety, minimizing the risk of getting scalded with excessively hot water.

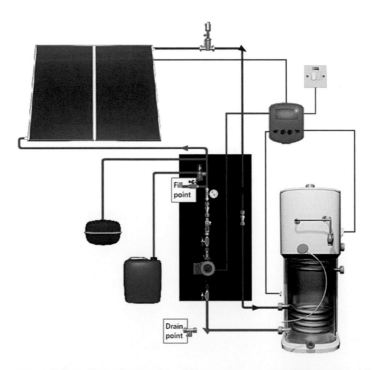

One of the objectives of a solar system controller is to maintain the user's safety

Types of controller

There are several types of controller, depending on how they measure and sense changes. The main one used in the UK is the Differential Temperature Controller, or DTC. However, there are at least two more ways of controlling the solar domestic hot water system which are not used in the UK.

Light

The pump can be powered by light intensity, using a photovoltaic cell to supply electricity to the pump, or another method can be to use light intensity sensors to activate pump circulation. Neither type of light intensity control is allowed in the UK. This is because light intensity controllers can, in certain scenarios, pump heat out from the collector and so the system is not 'interlocked'.

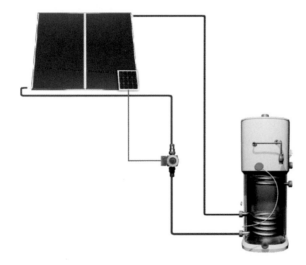

Light intensity can be used to power the pump

Gravity

Gravity, or thermosiphon primary circuits, are found in the Mediterranean region and use natural circulation controlled by the Sun, therefore no pump or electricity is needed. However, it is less efficient under conditions of low radiation and can give rise to very hot water in the storage tank. It must also be built to avoid air-locks, therefore the storage tank needs to be higher than the collector. This type of primary circuit control is not typically used in the UK.

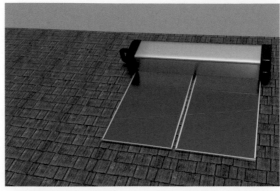

Gravity circuits are found in the Mediterranean regions and use natural circulation

E-LEARNING

Use the e-learning programme to learn more about DTCs.

ACTIVITY 31

Thinking about a thermosiphon solar system, is it interlocked (i.e. back up heating cannot be pumped out to the ambient environment) and is it a form of differential temperature controller?

Solar primary circuit controls There are three types of controls – Differential Temperature Controller (DTC) or light intensity sensors, Photovoltaic cells to power the pump, or thermosiphon primary circuits which use gravity – not typically used in UK

Issues to consider with solar primary circuit controls

UK AND INTERNATIONAL STANDARDS

Whenever installing any electrical components of the solar domestic hot water system remember to use Part P of the Building Regulations, particularly in wet or damp areas such as kitchens and bathrooms. Like all standards, building regulations are subject to change. Your Trade Association or solar supplier should keep you up to date with the latest standards and regulations.

There are a number of issues to consider with solar primary circuit controls

HEALTH AND SAFETY

Part P, electrical safety of the building regulations applies in England and Wales. All European countries have their own version of the building regulations or standards and it is vital that the local, regional and national electrical and construction procedures and requirements are followed. This allows another electrical fitter to work on the circuit and know what specification has been used for that job. It is also important that as well as following the appropriate building regulations, the manufacturer's instructions are also followed so that the system works as designed and a maintenance or other installer can have further information on the built system.

DIFFERENTIAL TEMPERATURE CONTROLLER (DTC)

Purpose of a differential temperature controller

To efficiently control a system circulating pump within the hot water system the primary circuit control should operate as efficiently as possible, wasting no solar energy, but equally not absorbing too much so that safety is compromised. When solar radiation is available and required by the storage vessel the controller should economically activate the circulating pump to transfer heat from the collector.

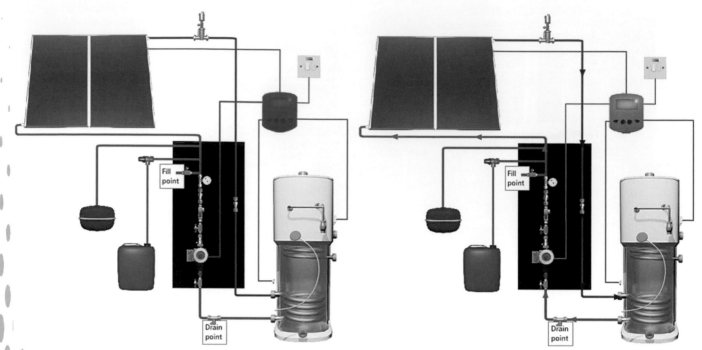

To start the pump when solar energy is available and can be used in the storage vessel

When solar radiation is not available or not required when the storage vessel has reached the required temperature, the controller should prevent unnecessary use of electricity by the pump and potential 'heat export' from the storage vessel to the collector.

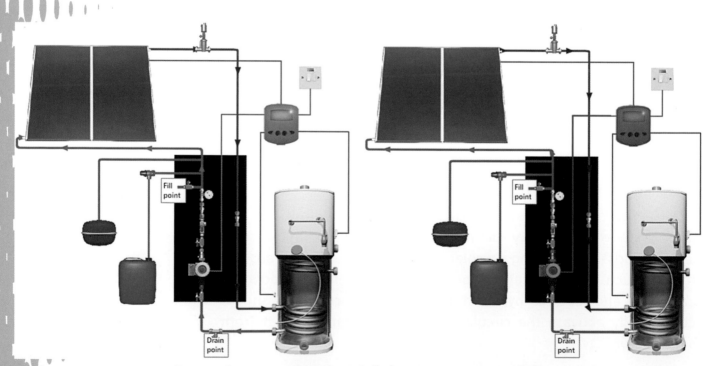

To stop the pump when there is little storage energy available

HEALTH AND SAFETY

The controller should only allow sufficient solar energy transfer to the storage vessel as is useful but safe. Unrestricted heat transfer could result in dangerously high hot water temperatures at taps and failure of hot water appliances operating outside their design specification (e.g. rubber and plastic components in mixer showers).

To efficiently control a circulating pump within the solar domestic hot water system the DTC measures temperatures at two or more points and compares these values against preset or programmable switching values. It also has a thermostatically set maximum hot water temperature.

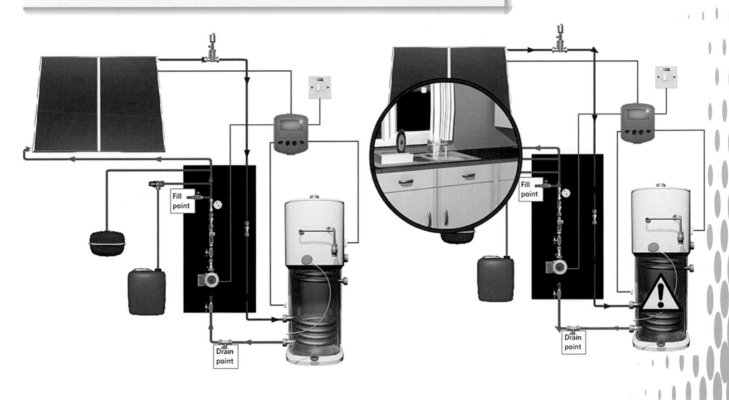

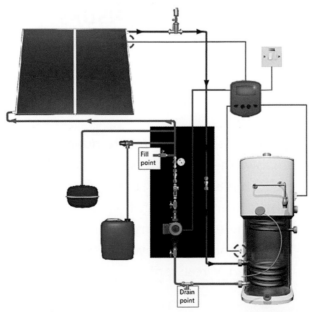

To prevent the hot water in the storage vessel overheating and becoming unsafe

E-LEARNING

Use the e-learning programme to see a demonstration of a Differential Temperature Controller.

ACTIVITY 32

In a typical property with solar heating, the central heating control system provides time and temperature control of the space and back up hot water heating, and the separate DTC controls the on/off and thermostatic temperature of the solar system. These are normally two separate control circuits. Can you see any benefits with combining these two systems into one centralized programmer?

TEMPERATURE SENSORS

There are two key control sensors; one measuring temperature at the highest point in the collector, the other at the lowest point in the storage vessel, between or just above the solar heat exchanger connections.

The sensor measuring temperature at the collector must perform reliably, therefore the sensor head is made from stainless steel and the cable sheathings made from silicone rubber. This protects the sensor from effects of very high collector temperatures and UV light. The external sensor has black sheathing in the figure opposite. The external cabling which is attached to the silicone sensor cable must be UV resistant.

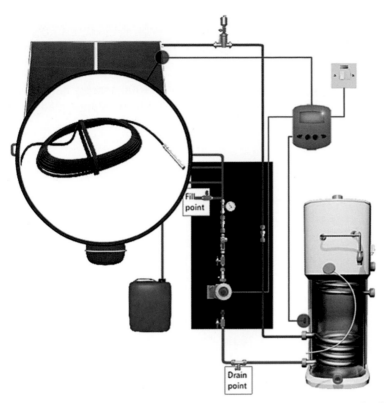

Two key controllers; the first, measuring temperature at its highest point, the second, at it's lowest point

A third sensor is normally located near the top of the storage vessel, to record the delivery temperature of the hot water to the taps.

Where there are split collector primary circuits, an additional sensor is needed on the second solar collector. Again this needs to be a sensor and cabling with high temperature and UV protection.

Third sensor, located near top of storage vessel

ACTIVITY 33

In the table below, the resistance values of PT1000 sensors are shown. What temperature would the sensor be reading if the resistance reading on your multimeter is:

1. 1097 ohms
2. 1366 ohms
3. Infinity ohms

°C	°F	Ω		°C	°F	Ω
−10	14	961		55	131	1213
−5	23	980		60	140	1232
0	32	1000		65	149	1252
5	41	1019		70	158	1271
10	50	1039		75	167	1290
15	59	1058		80	176	1309
20	68	1078		85	185	1328
25	77	1097		90	194	1347
30	86	1117		95	203	1366
35	95	1136		100	212	1385
40	104	1155		105	221	1404
45	113	1175		110	230	1423
50	122	1194		115	239	1442
Resistance values of the PT1000 sensors						

Sensor cabling sizes

The resistance of the sensor is affected by the cabling, so it is important that the right size of cable is used, and this is determined by the length of cable required. Wiring run lengths and cross sectional areas for low voltage wiring applies, so for lengths up to 50m the cross sectional area should be 0.75mm², and for lengths over 50m but less that 100m the cross sectional area should be 1.5mm².

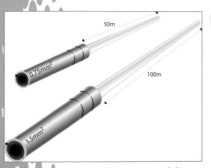

Different sizes of sensor cabling

ACTIVITY 34

You go to inspect a house and find out that telephone wiring cable has been used to connect the solar sensor to the DTC. What advice would you give to the customer? What effect will telephone cable probably have on the readings?

DTC OPERATION

Operation of DTC

The DTC operates by monitoring the three key temperatures via the sensors and turning the pump on or off as appropriate. It requires temperatures to be set to activate and deactivate the pump to make an efficient and safe solar hot water system. These key temperatures are:

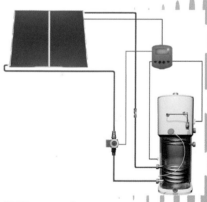

- 'T on'
- 'T off'
- 'T max'

DTC operating

T on

'T on' is the temperature that activates the pump circulation. If the value is seven then the collector (hottest) would need to be more than 7°C hotter than the bottom of the storage vessel (coolest) before system circulation begins. This ensures fluid is only circulated when there is energy to be collected from the panel.

T on is the temperature that activates the pump circulation

T off

'T off' is the temperature difference that deactivates pump circulation. If the value is four then the collector (hottest) would need to be less than 4°C cooler than the bottom of the storage vessel (coolest) before system circulation stops.

There must be a difference between 'T on' and 'T off' to prevent the circulation pump from cycling on and off rapidly, which will significantly reduce the energy collected. The values 'T on' and 'T off' should also be increased for primary circuits with longer pipe runs and greater fluid volumes.

T off is the temperature difference that deactivates pump circulation

T max

'T max' is the maximum temperature allowed in the storage vessel, which is normally set at 60°C in order to meet Legionella sterilization and prevent dangerously hot water in the taps. When a solar circuit reaches this temperature the pump will be switched off.

T max is the maximum temperature allowed in the storage vessel

ACTIVITY 35

A DTC is set to T_{on} 7°C and T_{off} 4°C and T_{max} 60°C. What will be happening at each of the following situations?

1. The solar section of the cylinder is at 55°C and the collector is at 63°C
2. The solar section of the cylinder is now at 59°C and the collector is at 62.9°C
3. The solar section of the cylinder is still at 59°C and the collector is at 66.1°C
4. The solar section of the cylinder is now at 60°C and the collector is at 67.1°C
5. The customer opens the tap and the collector is at 85°C and the cylinder drops to 55°C

CHECK YOUR KNOWLEDGE

1. **On the drainback primary circuit shown, three sensors need to be placed. On the diagram indicate where they should be placed.**

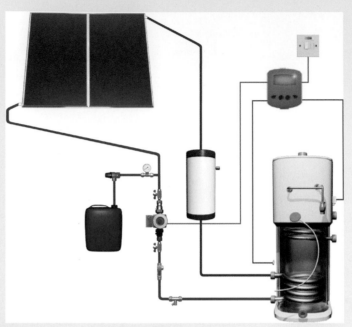

2. **Match up the settings shown with the relevant pump action.**

Action	Setting	Action
Activates the pump when the temperature between the collector and the top of the storage vessel is more than a set value	T on	Deactivates the pump when the temperature between the collector and the top of the storage vessel is less than a set value
Activates the pump when the temperature between the collector and the bottom of the storage vessel is more than a set value	T off	Deactivates the pump when the storage tank reaches maximum temperature

Deactivates the pump when the temperature between the collector and the bottom of the storage vessel is less than a set value	**T max**	Deactivates the pump when the temperature between the top and bottom of the storage tank reaches the same temperature

Chapter 7

INSTALLATION MATERIALS AND FITTINGS

LEARNING OBJECTIVES

By the end of this chapter you will be able to:

- Identify the components in the solar domestic hot water system

- Understand the criteria for choosing each one

CRITERIA TO USE FOR SELECTION OF COMPONENTS

Components and criteria for selection

We have already looked at the collectors, storage vessels, primary circuit designs and controls. When selecting materials for a solar system all components need to be durable. They must be able to withstand temperatures above 150°C without distortion or failure – or at least be protected from the heat if they cannot; they must be fit for use, and not degrade under UV light if installed externally.

When selecting materials for a solar system they must be durable

ACTIVITY 36

In engineering terms, every component in every engineered system has a particular specification to fulfil the particular requirements of that location in the circuit. Thinking about this, can you think of any categories for grouping solar components? Here are some words in no particular order that might help with this discussion: external, secondary, flow line, return line, primary, flow control, safety, control. Please note that these words don't need to be used and you can use other concepts as well; the idea of this exercise is just to ask you to think about categorization and function. Please also note that you could write a book on this activity and whilst we are very pleased if you enter into this discussion in some detail, we have provided an example answer in the answer section and are just asking you to discuss the *flow line of the primary circuit*, otherwise you could be here for a very long time.

PIPES AND FITTINGS

Pipes

Pipes are used to connect the solar hot water system together and complete the circuit for the transfer fluid to flow round. Pipes are available in various materials but they must meet the criteria for durability. Pipes can be made of copper or stainless steel. Unless explicitly allowed by the system supplier, do not use plastic pipes or barrier pipes.

Copper tube is suitable for use in fully filled pressurized and drain-back systems.

Copper tubes

Stainless steel piping is available as a flexible tube suitable for use in fully filled pressurized systems. It is available in a range of lengths and diameters and with pre-insulated options. It is also available as individual or paired pipes, for flow and return. The ends can be formed using a forming hammer or forming bolt for use with BSP thread fittings.

Stainless steel pipes are suitable for fully filled pressurized systems

Plastic pipe or barrier pipe is not high temperature tolerant and not suitable for solar primary systems. It melts at high temperatures! Some suppliers use a special silicone pipe and this is an unusual exception.

Plastic pipes cannot withstand high temperatures

Fittings

Fittings can be compression fittings or solder fittings. Compression fittings are available with copper or brass olives but only brass olives can be used in solar primary circuits up to three BAR pressures. If you expect pressures higher than three BAR, brass pipe reinforcing inserts must be used. An important thing to note is that all fittings used within 2m of the collector must always be compression fittings.

Fittings can be compression fittings or solder fittings

Solder is commonly used but it must be of the correct type, other-wise it will fall off or melt! It needs to be used sparingly as well. Most lead-free solders have a melting point over 215°C and typically, flat plate collectors will not get close to this temperature. Silver solder is usually required for the higher temperatures that typically occur with tube collectors.

Lead solder melts around 185°C and stagnation temperatures can be higher than this with both types of collector; therefore lead solder will just melt and fall off! As before, do not use plastic fittings or fit-tings containing plastic components as they are not capable of with-standing high temperatures.

Remember to always follow the collector manufacturer's instructions.

When using solder it must be of the correct type - otherwise it will fall off/melt

ACTIVITY 37

Most manufacturers now recommend only brazing, compression joints or special track pipe fittings on solar circuits. A few years ago, lead-free solder and feed joints were more commonly used on solar primary circuits. Why has there been a trend away from lead-free solder joints?

INSULATION

Types of insulation

There are different types of insulation available but not all are suitable. Pipe insulation must be resistant to high temperatures and if installed externally, also UV resistant. But additionally it must also be pest resistant to bird or rodent attack. It's amazing what animals can eat!

Pipe insulation must resist high temperatures

Elastomeric insulation This type of insulation for pipes is not high temperature tolerant or UV resistant, although it is water resistant. HT elastomeric insulation is high temperature tolerant, water and UV resistant and should be used instead

Polyethylene insulation is not solar temperature resistant, it melts at these temperatures. Also, some of the **elastomeric insulations** available are not heat resistant to the temperatures experienced in the solar heating systems. You must always check the manufacturer's specification.

High temperature specification elastomeric insulation is solar temperature tolerant, water and UV resistant and is available with a range of protective outer layers such as braiding that can be used. It is also available as pre-insulation on stainless steel flexible pipe and is available in short lengths, which makes things a bit more convenient.

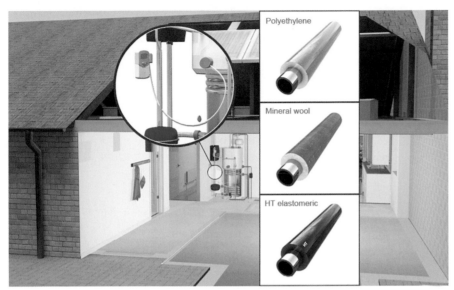

Always check the manufacturer's specification

ACTIVITY 38

If polyethylene insulation is not solar temperature resistant, where could it be used on the overall solar circuit?

Mineral wool insulation

Foil faced mineral or glass wool is excellent insulation for high temperatures but it is not weather resistant and is rarely used in domestic solar heating systems. It requires additional protection whenever used externally or in 'wet rooms'.

Foil faced mineral wool, with the added protection, is very high temperature and UV resistant and commonly specified for industrial tasks.

Foil faced mineral or glass wool is excellent insulation for high temperatures

ACTIVITY 39

Where would foil-faced mineral or glass wool insulation typically be specified as a requirement as compared to HT elastomeric insulation? What are the advantages and disadvantages of each material?

EXPANSION AND DRAINBACK VESSELS

Expansion vessels

An expansion vessel is part of a fully filled pressurized primary circuit and enables essential expansion and contraction to take place safely as the circuit heats up and cools down during the day and night. The performance of the vessel relies on a pressure relationship between a gas (air or nitrogen) and the heat transfer fluid separated by a rubber membrane. Unfortunately, the flexible rubber membrane that enables this to happen is only suitable up to a temperature of about 100°C.

The rubber membrane must be protected from heat above 100°C at all times. One way of doing that is to have a long leg with sufficient capacity to hold the equivalent of the collector fluid. The long leg

An expansion vessel is part of a fully filled pressurized primary circuit

should also have a downwards section to the expansion vessel to reduce 'heat creep' from the solar return pipe. It is important to have the vessel positioned on the return system and mounted downwards.

The second way of protecting the membrane is to have a large enough expansion vessel which allows it to hold the equivalent collector fluid volume under conditions of no solar radiation, when it is dark! Then at stagnation when the hotter fluid is driven into the expansion vessel by steam it mixes with the cooler fluid already held, which cools it down and so protects the membrane.

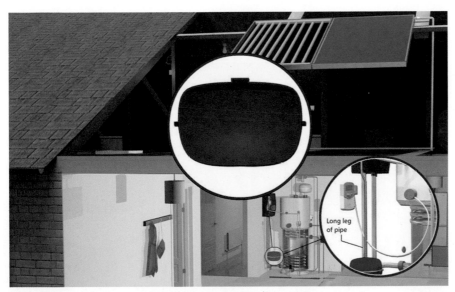

Should have a downwards section to the expansion vessel

E-LEARNING

Use the e-learning programme to see a demonstration of expansion vessels.

Expansion vessel membrane

The expansion vessel is delivered with the membrane pushed against the upper half, filled with a gas at an equal or greater pre-charge pressure than the system fluid pressure.

The solar system is filled to allow a 'thermal buffer' of fluid to fill the vessel with an equivalent volume to the collector fluid content. This is done before any solar heating or circulation takes place.

When the system is heated up or stagnation has occurred and steam fills the system, the transfer fluid is moved to the expansion vessel, mixes with the cool fluid and the mean temperature is reduced and the rubber membrane protected.

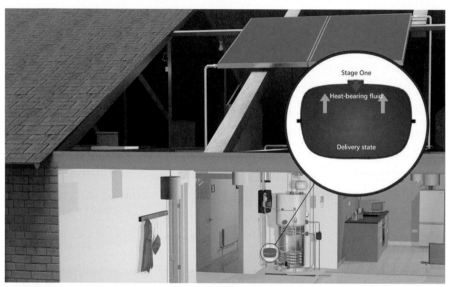

Stage 1

Stage 2

Stage 3

ACTIVITY 40

As solar thermal systems develop, expansion vessel manufacturers are developing higher performance expansion vessels and it is now common to find vessels rated up to 130°C. In the previous text, we have listed 100°C as a typical rated value of a solar expansion vessel. There is some fairly complex maths to very accurately size expansion vessels for solar thermal systems. This sizing is set out and freely available on the Web in a paper by Hausner and Fink called 'Stagnation behaviour of Solar Thermal Systems'. Please answer the following questions:

1. Why have we used a conservative figure of 100°C in the text above?
2. Can you provide (or have a think how you might provide) a simple rather than complex method for sizing solar expansion vessels?
3. In the Hausner and Fink paper, they discuss good and poor emptying behaviour of collectors. What do you think might be classed as good and poor emptying behaviour?

Drainback vessels

The position of the drainback vessel is carefully worked out and the volume of fluid is also carefully calculated to ensure efficiency in the system. Drainback systems rely on air present in the system to allow for the fluid expansion and drainback principle to occur when circulation is stopped – if incorrectly fitted, this can make them a little noisy!

Drainback vessels are specified for flow or return pipe connection and constructed to meet the highest temperature experienced in that part of the primary circuit, so always check the manufacturer's instructions when selecting them.

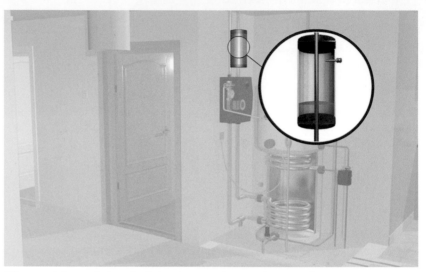

The position of the drainback vessel must be carefully worked out

When circulation is taking place very little fluid is in the drainback vessel, but there is lots of air. When circulation stops the fluid collects in the drainback vessel; this is all the fluid in the collector and the pipework above the level of the vessel itself. If it is a sealed system the air in the vessel is compressed slightly when the circulation fluid expands.

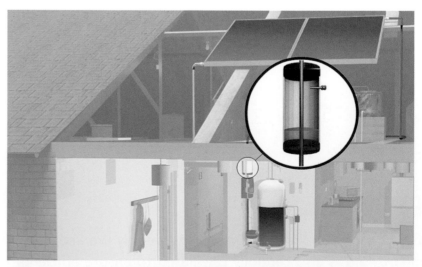

When circulation is taking place little fluid but lots of air is in the drainback vessel

The amount of fluid in the drainback primary circuit needs to be correct for that installation. Too much transfer fluid and it takes longer to heat up and circulate through the circuit.

Not enough fluid and it cannot fill the pipework to circulate all the way round and therefore no heat transfer can take place.

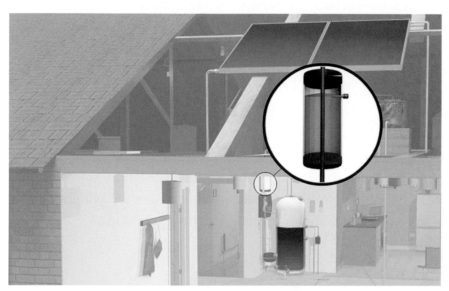

Too much transfer fluid may cause it to take longer to heat up

E-LEARNING

Use the e-learning programme to see a demonstration of drainback vessels.

ACTIVITY 41

Please state how you would calculate the drainback vessel size:

AIR VENTS AND SEPARATORS

Air vents and separators are designed to rapidly remove air from fully pressurized solar primary circuits by allowing air to rise in the separator and escape via manual or automatic air vents. As the heat transfer

Air vents and separators

fluid enters the chamber it slows down temporarily, allowing air bubbles to rise to the top where there is a manual and/or automatic air vent. It must be installed correctly in order to work – with the vents on top! The air vents must be high temperature tolerant – brass air vents used in central heating have a plastic float which would melt if used in a solar system, so must be solar quality.

After commissioning, automatic air vents are isolated to prevent pressure loss during stagnation, when transfer fluid could escape as steam through an automatic air vent.

Air vents are always installed vertically and separators are normally horizontal in a system but can be vertical. If installed vertically, positioning above a pipe with the flow moving towards the separator is efficient. If they are upright the flow in the pipes below them must be towards the vent, to carry the air bubbles. If the flow is away from the separator few air bubbles will escape because they will be carried by the flow – a bit like trying to go up a down escalator!

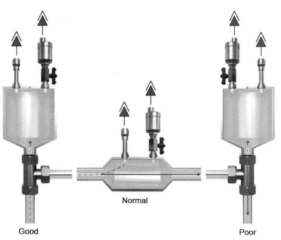

Good   Normal   Poor

Air vents installed vertically; separators installed horizontally, but can be vertical

ACTIVITY 42

Where should an air vent and separator be fitted on a solar circuit? What is the commissioning action related to solar air vents and separators?

PUMP STATIONS AND FLOWMETERS

Pump stations are usually bought with a combination of necessary components pre-assembled for rapid installation. All the standard components are included, all are solar rated and the electrical components are pre-wired. This makes installation much easier and quicker. The pump station will have removable insulation panels so it can be viewed during installation and maintenance.

Standard components are:

- Circulation pump
- Pump isolating valves
- Check valve(s)
- Pressure gauge
- Safety relief valve
- Flow regulator
- Expansion vessel connection
- Thermometer(s)
- Fill point

Optional components include:

- DTC
- Air separator

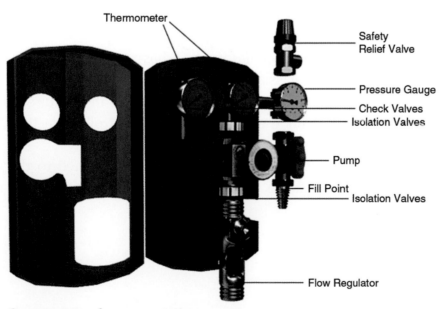

Components of a pump station

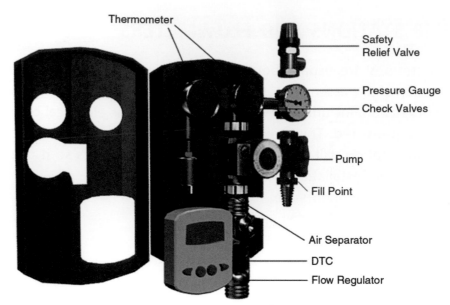

Optional components; DTC, air separator

Flow regulators

Flow regulators control the flow in the primary circuit and there is an optimum rate of flow for heat transfer fluid passing through the solar collector. The collector manufacturer defines the optimum rate and it needs to be checked that the installed system delivers this flow rate.

Flow rate can be controlled in three ways:
The first is via the speed settings on the circulation pump. This should be set to the lowest setting to reach the collector manufacturer's recommended flow rate.

The second is by a flow regulator which has a mechanical restrictor with a measured sight glass and adjustable throttle. After setting the pump speed, this restrictor should be adjusted to the required flow rate. The third way is for the flow rate to be controlled via the DTC and electrical supply to the pump.

Some solar controllers modulate the electricity supplied to the circulation pump. This feature should only be used if specified by the collector manufacturer.

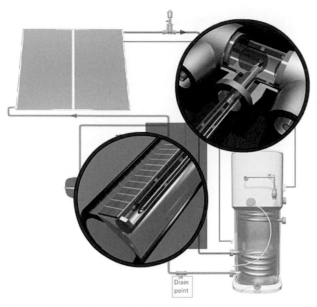

Rate of flow through the solar collector

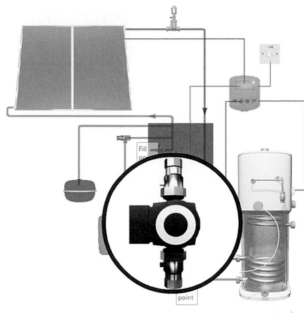

Circulation pump should be set to it is lowest setting

Flow regulator can control the rate of flow

ACTIVITY 43

Many suppliers recommend flow rates of around 1 litre/minute/m^2 collector. If this is the case, what would be the recommended flow rate for a 4.2m^2 collector and what are the advantages and disadvantages of higher flow rates?

POSITIONING OF COMPONENTS

Positioning criteria to follow when installing

UK AND INTERNATIONAL STANDARDS

Positioning the components within a solar primary circuit is important for the safe and efficient functioning of the system. Always follow the manufacturer's instructions. Always select solar grade components. Always follow BS 7671 when cabling.

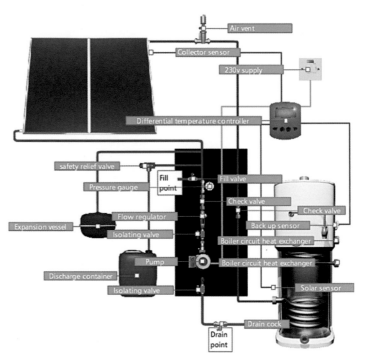

Must follow a certain positioning criteria when installing

Special care is needed when cables could come into contact with hot pipes or hot water cylinders and thermal stores. With solar circuits, cables also need to be UV resistant and watertight where they are outside. Special care also needs to be taken in bathroom and kitchen areas. Sensors need to be attached very securely to prevent accidental removal – which would affect the system operating correctly if they accidentally became loose.

A good practice is to label power and sensor cables wherever they are close together or look similar. Also, make sure all sensor cables have the correct diameter to avoid resistance losses which can affect the solar controller.

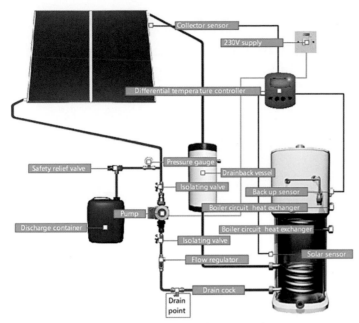

Cables must be UV resistant and watertight

Fully filled pressurized solar primary circuit

Before we leave this chapter have another look at a fully filled solar primary circuit to refresh your memory.

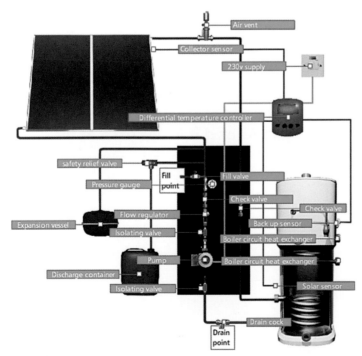

Fully filled solar primary circuit

Drainback solar primary circuit

Before we leave this chapter have another look at a drainback solar primary circuit to refresh your memory.

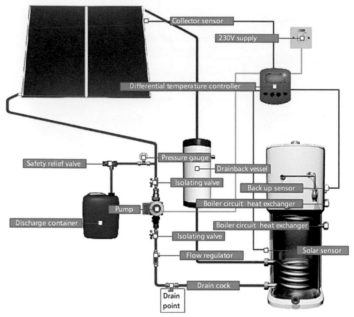

Drainback solar primary circuit

CHECK YOUR KNOWLEDGE

1. **Complete the table shown by selecting the type of pipe that can be used in fully filled pressurized systems and drainback systems or if it is not used at all.**

Type of pipe	Fully filled pressurized solar circuits	Drainback solar circuits	Not typically used in solar circuits
Plastic			
Copper			
Stainless steel			
Flexible stainless steel			
Barrier			

2. **Identify the numbered parts of the pump on the image shown.**

3. **A pre-assembled pump station has nine standard components and can have at least two additional items. Identify the nine standard items you would expect and two additional items.**

Component	Standard	Additional	Not Applicable
Air separator			
Check valve(s)			
Circulation pump			
Drain point			
DTC			
Expansion vessel connection			
Expansion vessel			
Fill point			
Flow regulator			
Heat exchange coil			
Pump isolation valves			
Pressure gauge			
Safety relief valve			
Thermometer(s)			

4. Which of the following items must not (unless manufacturer states otherwise) be used in solar domestic hot water systems?

☐ a. Compression fittings with brass olives

☐ b. Lead solder fittings

☐ c. Silver solder fittings

☐ d. Plastic fittings

☐ e. Polyethylene insulation

☐ f. HT elastomeric insulations

5. How are the rubber membranes on expansion vessels protected from the heat? Select the options that you think are correct.

☐ a. Positioned at the end of a downward pipe

☐ b. Positioned at the end of a long leg of un-insulated pipework

☐ c. Supplied with check valves leading to the vessel

☐ d. Putting nitrogen gas in the lower part of the vessel

Chapter 8

FILLING, COMMISSIONING AND MAINTENANCE

LEARNING OBJECTIVES

By the end of this chapter you will be able to:

- Identify Health and Safety considerations when filling and commissioning a solar hot water system

- Identify installation checks to be made prior to commissioning a system

- Describe hand filling methods

- Identify the process to complete when commissioning a solar hot water system

- Describe documentation and handover of the system to the customer (user)

- List maintenance procedures and tests

- List possible faults and their rectification for pressurized and drainback systems

Filling, commissioning and maintaining a solar hot water system

FILLING AND COMMISSIONING

The most important thing to establish is that the installed system is leakproof and works well. This might involve bypassing some of the check valves while filling, then, when the system has been tested as OK and is commissioned for use, some of the valves and settings need to be changed again. This could be confusing! Therefore extensive use is made of checklists during the filling, commissioning and maintenance process in order to reduce risk.

HEALTH AND SAFETY

Safety precautions also need to be carefully applied on a new system, particularly on hot sunny days, as some of the pipework could be hot, even though it would be protected from the Sun.

The installed system must be leakproof and function well

ACTIVITY 44

Bearing in mind the stagnation temperature of solar collectors, are there any special precautions that should be taken into account when installing collectors when there is reasonable daylight or sunlight? What methods can be used to avoid steam being generated in the solar primary circuit during filling?

Installation checks

Installation checks are made to ensure the installed system first complies with the original design specification for the customer, and then checked for any obvious defects.

> **Installation checks**
> Are made to ensure the installed system first complies with the original design specification for the customer, and then checked for any obvious defects

Installation checks ensure the system complies with original design specifications

Installation checklist

The inspection looks at and checks the items listed on the checklist:

● All pipes are well supported (no sagging) particularly on a drainback circuit

- Hydraulic connections are secure, in preparation for leak testing

- Primary circuit valves are in the 'filling' positions (especially check valves and mechanical isolator on AAV)

- Electric cables are secured (using clips or conduits) and are the correct size (to carry current or minimize resistance)

- Power and sensor cables are different in appearance or identifiable and labelled when required

- Electrical supply for a solar system is properly fused and labelled (it should be a three amp fuse for 240 volt supply)

- All earthing and bonding for the installation must be to BS 7671:2008

- Solar sensors are secured in correct positions

- Electrical connections are correct, cable terminations are tidy and secure

- Back up boiler cylinder thermostat is correctly positioned and secured; note that the thermostat should be set to about 60°C and should be positioned approximately $1/3$ the height of the boiler volume taken from the bottom of the boiler heat exchanger

- Collector fixings are secure and as per fitting instructions

- Solar collectors are covered before filling

- Roof covering made good and all tiles and collector mounting frames returned to correct positions

- Holes in building fabric made good, and fire, heat loss and weathering barriers maintained

- Pipe insulation applied to solar primary circuit pipes, hot water distribution pipes and boiler primary pipes leading from storage vessel, to be added AFTER leak tests have been carried out

- The safety relief valve is terminated either externally in a safe location or internally into an appropriate catchment vessel

- All solar primary components such as expansion vessels are correctly located, fixed and pressure set according to manufacturers' instructions

- All safety and information labels are in place.

Inspection must follow the items listed on the installation checklist

ACTIVITY 45

On a solar circuit, what is the correct location of the expansion vessel and how and when do you go about checking the pre-charge pressure? Are there any major items not mentioned on the above list and, thinking of the major item not mentioned, how would you check this item?

INITIAL FILLING

Filling is carried out in two stages; first there is an initial fill and then the main fill; each has a different purpose. The initial filling is to flush out the system and confirm the basic system functions work and there are no leaks. The main filling prepares the system for use.

During the initial fill you are preparing the system for use; checking the basic system functions work correctly, such as the pump, that the fluid actually flows through the pipes and is not blocked – or leaks! The fluid used is usually warm water – but always check the manufacturers' instructions as there are variations.

Two stages in filling; Initial filling, main fill

Filling methods

Filling solar primary circuits can be carried out in a number of ways; we will briefly look at two. However, it is important to know that filling loops used for central heating systems are NOT suitable for use with solar systems. This is because filling loops can lead to dilution of antifreeze or overfilling of drainback circuits.

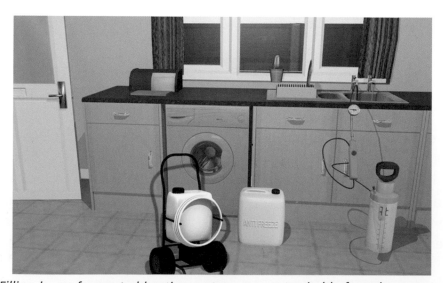

Filling loops for central heating systems are not suitable for solar systems

The hand filling method

Hand filling using a filling bottle can be used for both drainback systems and fully filled pressurized primary circuits. Fluid is slowly pumped into the system. A fully filled pressurized primary circuit is filled from a low point so that the air is pushed up towards the air vent at the top of the circuit. Hand filling can be simplified by bypassing check valves to avoid large pockets of trapped air, where there is a pump station fitted. A drainback system is filled from the filling point until fluid is seen at the 'fill level'.

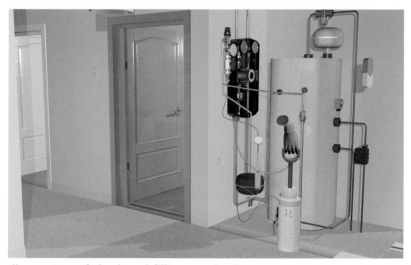

Illustration of the hand filling method

The machine filling method

A filling machine cannot be used for a drainback system, but can be used for a fully filled pressurized system. The fluid is pumped into the circuit via two filling points above and below an isolating valve and the air is pushed towards the lower filling point where air and fluid is passed into the reservoir of the filling machine, so the filling machine acts as an air separator.

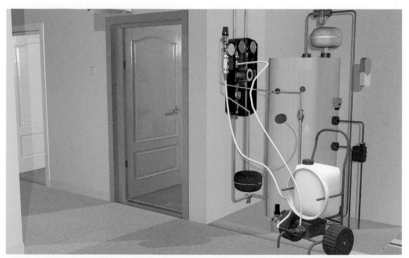

Filling machines can only be used for a fully pressurized system

E-LEARNING

Use the e-learning programme to see a demonstration on the filling methods.

ACTIVITY 46

It is stated here that machine filling is not appropriate for drainback solar systems. Why is this and how could machine filling be adapted for use on a drainback circuit?

Testing and flushing the system

The first thing to do once the system is filled is to test for leaks. To do this thoroughly the system needs to be put under pressure. This should be higher than normal operating pressure, and be near to the maximum operating pressure of the safety relief valve. If any leaks are found they must be repaired and the initial filling procedure repeated until there are no more leaks detected. It is not possible to carry out this test with pipe insulation in place!

Once the system is filled, must test for leaks

The initial filling is usually made with warm water, as this helps the removal of any excess flux, loose filings and debris within the system. The system needs to have these things removed to prevent blockages and improve circuit performance. In order to ensure as much as possible is removed you need to keep the pump activated for some time.

Warm water is used to remove excess flux, loose fillings and debris

ACTIVITY 47

What is the recommended testing pressure for a primary circuit and how would this test pressure be realized on a solar primary circuit? What would be the recommended test pressure with a three bar and a six bar safety relief valve?

Completing the initial fill

Once all the tests and checks have been made and the system is leak free, and the fluid flows freely without any debris causing blockages, the water is drained from the system. It is important to make sure all the water is removed from the system because any left will dilute the antifreeze used as the transfer fluid and reduce the effectiveness of the system.

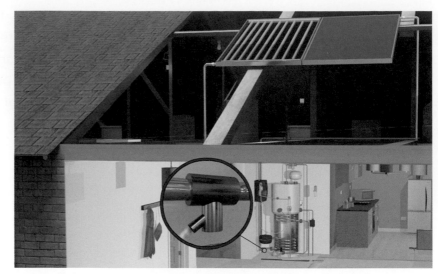

Once all the tests are complete the water must be drained

MAIN FILLING OF THE SYSTEM

Preparation for the main filling

The main filling completes the preparation of the system ready for use by the customer. At this point you can be confident that there are no leaks and the main components all work properly. All that needs to be done, in brief, is to fill the circuit with the correct fluid; check again all is working well by commissioning the system; then handover the documentation to the customer, showing them how to maximize the effectiveness of their new system and explain the maintenance process.

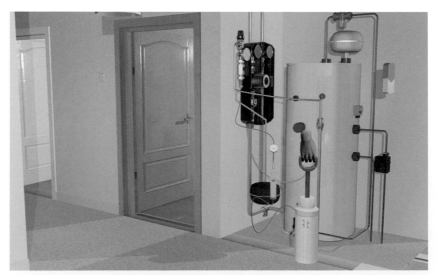

The filling completes the preparation, making it ready for the customer

UK AND INTERNATIONAL STANDARDS

All systems in the UK require antifreeze to be added to the transfer fluid, unless stipulated by the manufacturer. Some direct and drainback systems do not use any antifreeze but you should always check the manufacturer's instructions to be sure. The antifreeze used is Propylene Glycol, which has a low toxicity and neutral pH. A 50 per cent water/Propylene Glycol mix can protect down to −30°C so in the UK usually needs less than 50 per cent, but it will vary according to location, i.e. between Cornwall and John O'Groats.

Transfer fluid may be pre-mixed or can be mixed at the site – in which case it needs mixing early in order to allow any air bubbles to escape before you add it to the system.

HEALTH AND SAFETY

When handling fluids, always follow the COSHH guidelines.

Pre-mixed antifreeze solutions can be used straight from the container. They should not be diluted or their effectiveness will be reduced.

Always check manufacturer's instructions

ACTIVITY 48

What does COSHH stand for? How must the waste antifreeze be disposed of?

Filling a fully filled pressurized system

When filling fully filled pressurized systems the heat transfer fluid (as specified by the manufacturer) is added to the primary circuit to the correct operating pressure using an appropriate filling method, such as a hand pump or filling machine. If antifreeze has been added and you are uncertain of its strength, such as when some of the initial filling water may have remained in the system, the heat transfer fluid should be tested using a refractometer.

Again air must be purged from fully filled pressurized systems. To test the circulation, the system should be activated using the manual override function on the solar controller.

Refractometer Used to determine whether the heat transfer fluid has the correct concentration of antifreeze

Filling fully filled pressurized systems

Filling a drainback system

Filling a drainback system can only be carried out using a hand operated bottle method. The heat transfer fluid is added to the desired level – which is usually the overflow valve on the drainback vessel. Circulation should then be started using the manual override function on the solar controller. Air pockets are likely to be found and dislodged during this process and so you will probably need to top up to the correct level once the system has settled. Always check, usually by listening carefully, that when the solar controller is switched off, all the fluid drains back into the drainback vessel.

To fill a drainback system a hand operated bottle method must be used

ACTIVITY 49

Antifreeze tends to entrap air. Discuss how you might minimize the air in the antifreeze solution before filling the circuit.

Main fill summary

In summary, the main fill involves getting the heat transfer fluid to the correct strength specified by the manufacturer, which could be by purchasing a pre-mixed transfer fluid or mixing your own. When handling fluids, always follow the COSHH procedures. Circulating the fluid and removing air is most important during this phase. Fully filled pressurized systems need the air purged from the system, and drainback systems need to have the fluid level checked after it has been circulated in case of trapped air pockets. If you need to check the strength of the transfer fluid you can use a refractometer.

Filling the system

COMMISSIONING

Commissioning procedures involve checking and setting of system safety and operational functions ready for daily use. Use of a commissioning checklist at this point is useful to ensure nothing is missed out or forgotten.

Must check and set system safety and operational functions

Commissioning checklist

- System flow rate is set according to the manufacturer's instructions
- The pump speed should be set as low as possible to achieve the desired flow rate so as to minimize 'auxiliary' energy losses:

- Where the solar circulating pump is 240 volts a typical flow would be one litre per minute for each square metre of collector
- Where the solar circulating pump is PV driven, sometimes with less power available, a typical flow might be lower than a mains powered primary circuit

- Test the operation of the safety release valve and ensure the valve reseats and fluid discharge stops. If necessary, re-top up fluid volume

- Adjust any thermostatic mixing valves installed in the household hot water distribution pipes to a safe temperature

- Adjust back up heat source time and temperature controls (e.g. central heating control system) to maximize solar benefits, remove timed daylight boiler activity. Check 'Legionella' settings on back up control system

- Adjust and check solar controller settings, readings and functions:

  - T on
  - T off
  - T max
  - Sensor temperature readings
  - Other desired functions (e.g. activation of variable pump speed control)

- Check system pressure gauge is functioning correctly
- Check circulation is not noisy
- Set valves to normal operation; including isolation of automatic air vents and correct position of check valves
- Uncover solar collectors
- Ensure the system controller is left in 'automatic' mode

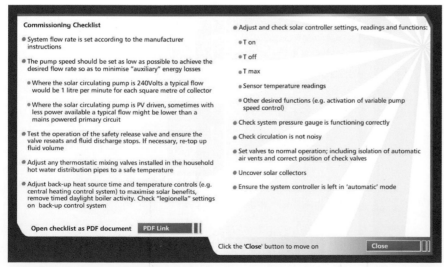

Commissioning checklist

ACTIVITY 50

The list discusses setting the circulating pump, if possible, on its lowest setting to minimize auxiliary losses. This would normally be setting 1 (of 3) on a typical 5m head central heating circulating pump. What would be a typical power rating in watts of a 5m head pump on setting 1 and how might you reduce this auxiliary loss?

Are there any useful actions to cover at the commissioning stage to improve the handover stage of the installation process?

Setting the system for use

The few final checks and adjustments complete the commissioning process. An important check is to make sure the system is not noisy. It should be very quiet in use as the fluid circulates in the system.

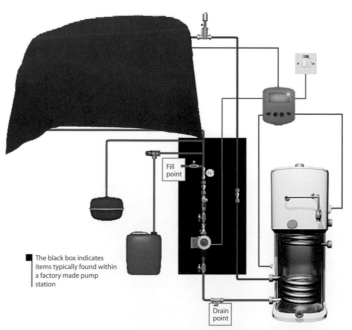

The black box indicates items typically found within a factory made pump station

Must ensure that the system is not noisy

Valves

Make a final check that all the valves are set for normal operation. This includes the isolation of automatic air vents and the correct position of the check valves.

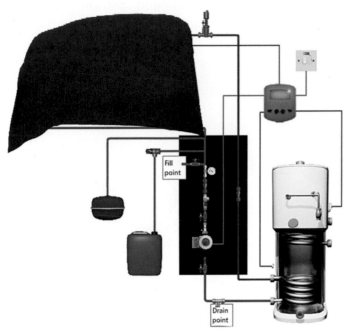

Ensure all valves are set for normal operation

Solar collectors

The next task is to uncover the solar collectors on the roof. These would have been covered since they were installed. If there is daylight, the system should now begin to warm up and start to heat the storage tank.

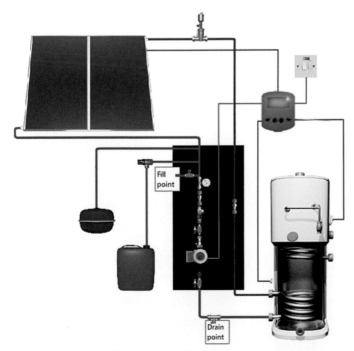

Uncover solar collectors on the roof

Controller

Make sure the system controller is set to automatic, which is how it should be left for the customer. Remember that if the installation is in a soft water area where limescale is not a problem, Tmax can be increased from 60°C to 75°C provided blending valves have been added into the system to prevent scalding.

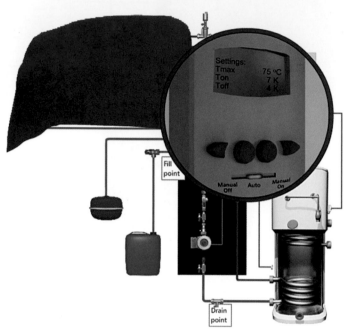

Ensure system collector is set to automatic

ACTIVITY 51

There is some discussion ongoing in the solar industry about using the DTC and its functionality to control the back up boiler (or immersion or heat pump, etc.) HW program. What two parameters would be required from the DTC to control the back up hot water circuit and which sensor would be required to enable this functionality?

HANDOVER TO THE CUSTOMER

The handover should leave the customer confident they can look after the system correctly, and obtain long trouble free use of solar energy for their hot water system. This essential element in the process is to leave them feeling confident that they know what to do to keep the system working at optimal efficiency and that they know what to do if things go wrong.

Handover to the customer

The customer will need to hold a number of documents and understand that they need to keep them safe, as they will be needed on maintenance visits or in case something goes wrong. The list is shown in the Documentation checklist.

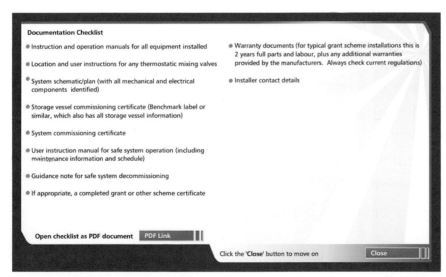

Documentation checklist

Documentation checklist

- Instruction and operation manuals for all equipment installed
- Location and user instructions for any thermostatic mixing valves

- System schematic/plan (with all mechanical and electrical components identified)
- Storage vessel commissioning certificate (benchmark label or similar, which also has all storage vessel information)
- System commissioning certificate
- User instruction manual for safe system operation (including maintenance information and schedule)
- Guidance note for safe system decommissioning
- If appropriate, a completed grant or other scheme certificate
- Warranty documents (for typical grant scheme installations this is 2 years full parts and labour, plus any additional warranties provided by the manufacturers. Always check current regulations)
- Installer contact details
- Any other documentation thought appropriate by the installation company, equipment manufacturer or required by building regulations, standards, accreditation or local authorities

User information

Along with all the documentation it is important that the customer holds a record of certain measurements for their installed system. The following measurements need to be recorded for the customer's installed system:

- System flow rate and pump speed
- Heat transfer fluid type, frost protection level, and total fluid content
- Solar collector net surface area
- Safety relief valve discharge pressure
- System operating pressure (for fully filled pressurized systems only)
- Expansion vessel gas pre-charge pressure (for fully filled pressurized systems only)

Optimum use of the central heating controls should be discussed with the customer. For Legionella risk minimization, they must be set to heat the back up volume of the cylinder to 60°C for an hour once a day. Depending on the customer's lifestyle and hot water requirements,

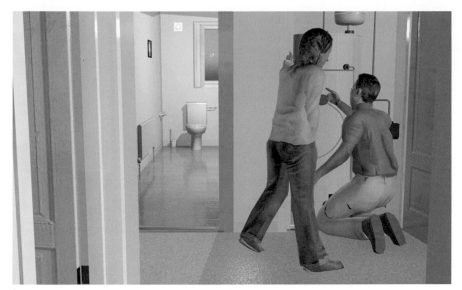

The customer must have a record of certain measurements

ideally, this hour should be set to early evening so that the most is used of any solar gain during daylight hours. After morning bathroom usage, leaving a cylinder full of lukewarm water for the Sun to heat up during the day maximizes the benefits of the solar system.

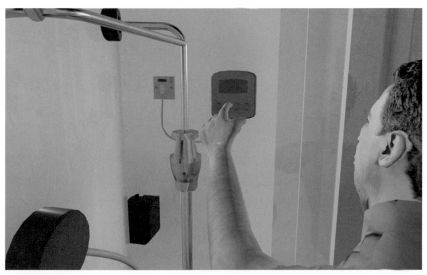

Optimum use of central heating controls should be discussed

The customer will need to look after the system between maintenance visits. Such tasks are all described in the user manual, however it is good practice to also explain the tasks, or even show the customer what to do in addition to leaving the user manual and documentation. The customer also needs to know when to call for help and recognize when something has gone wrong. For example, the customer should call out the installer if the safety relief valve is dripping or has released fluid.

Customer must look after the system between maintenance visits

ACTIVITY 52

Thinking about the DTC, what settings and information should the house-holder be able to access and what information is better to keep hidden from the customer's view?

Maintenance schedule for the customer

The final item to make sure the customer has and understands is the maintenance schedule, outlined with the intervals between the visits. As you complete the handover to the customer explain to them how to make sure they get the maximum benefit from their solar hot

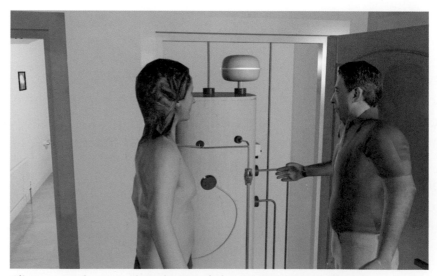

Client must have and understand the maintenance schedule

water system. This would include the optimum central heating settings to maximize solar performance.

MAINTENANCE

Maintenance visits are made at set intervals when the system can be serviced. These intervals can be between one and five years, with five years at the maximum limit.

The maintenance checklist gives the tests and checks to be carried out. Certain pieces of equipment will be needed for service visits – apart from normal toolkits. You will need some litmus paper to test the pH of the transfer fluid – which should be neutral, and a refractometer to measure the frost protection given by the transfer fluid, which should be the same as the original level in the documentation.

Litmus paper To test for pH of transfer fluid (it should be neutral)

Litmus paper is needed to test the pH of the transfer fluid

Maintenance checklist

- Collector glazing is damage free and clean
- Collector weather seals are sound
- Collector fixings are in good condition and secure
- Collector sensor is in position and secure
- External pipe insulation is in good condition
- Internal pipe insulation is in good condition
- Storage vessel sensors are in position and secure
- Solar primary circuit is leak free
- System pressure is correct (for fully filled pressurized primary circuits) and fluid level is correct (for drainback circuits)
- System flow rate is correct
- Circulating pump is at correct speed setting
- System circulation is not noisy
- Safety relief valve is operational and reseats correctly
- System controller functions are operational

- Cables are in good condition and system is correctly fused and earthed
- Heat transfer fluid is not acidic (pH test required)
- Heat transfer fluid can still protect against the frost (refractometer required)
- System components do not show signs of corrosion
- Everything is in working order and the solar system is performing well and producing lots of solar power

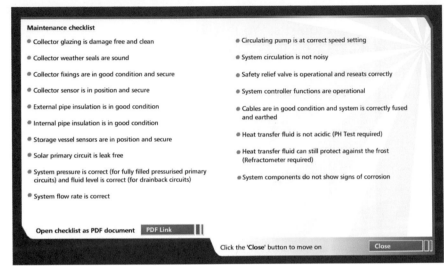

Maintenance checklist

ACTIVITY 53

Thinking about the maintenance checklist, what would you look for on a solar system to check for adequate performance?

FAULT-FINDING

Occasionally things go wrong and the system develops a fault. There are fault-finding checklists to help identify what could have gone wrong with a system. It can then be repaired and the system restored to proper functionality.

There are fault-finding checklists to follow

Many items are the same for each type of system; however the Fully Filled Pressurized System has a wider range of possible faults because it is a more complex system. The Drainback System only has a few faults that are unique.

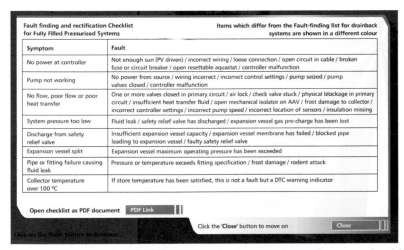

Fully filled pressurized systems fault-finding and rectification checklist

Fault finding and rectification Checklist for Fully Filled Pressurised Systems	Items which differ from the Fault-finding list for drainback systems are shown in a different colour
Symptom	**Fault**
No power at controller	Not enough sun (PV driven) / incorrect wiring / loose connection / open circuit in cable / broken fuse or circuit breaker / open resettable aquastat / controller malfunction
Pump not working	No power from source / wiring incorrect / incorrect control settings / pump seized / pump valves closed / controller malfunction
No flow, poor flow or poor heat transfer	One or more valves closed in primary circuit / air lock / check valve stuck / physical blockage in primary circuit / insufficient heat transfer fluid / open mechanical isolator on AAV / frost damage to collector / incorrect controller settings / incorrect pump speed / incorrect location of sensors / insulation missing
System pressure too low	Fluid leak / safety relief valve has discharged / expansion vessel gas pre-charge has been lost
Discharge from safety relief valve	Insufficient expansion vessel capacity / expansion vessel membrane has failed / blocked pipe leading to expansion vessel / faulty safety relief valve
Expansion vessel split	Expansion vessel maximum operating pressure has been exceeded
Pipe or fitting failure causing fluid leak	Pressure or temperature exceeds fitting specification / frost damage / rodent attack
Collector temperature over 100 °C	If store temperature has been satisfied, this is not a fault but a DTC warning indicator

Open checklist as PDF document   PDF Link

Click the **'Close'** button to move on   Close

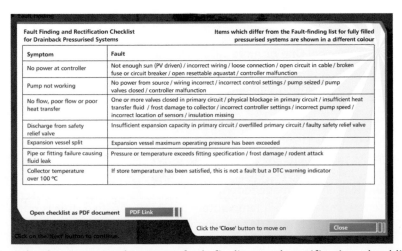

Drainback pressurized systems fault-finding and rectification checklist

Fault Finding and Rectification Checklist for Drainback Pressurised Systems	Items which differ from the Fault-finding list for fully filled pressurised systems are shown in a different colour
Symptom	**Fault**
No power at controller	Not enough sun (PV driven) / incorrect wiring / loose connection / open circuit in cable / broken fuse or circuit breaker / open resettable aquastat / controller malfunction
Pump not working	No power from source / wiring incorrect / incorrect control settings / pump seized / pump valves closed / controller malfunction
No flow, poor flow or poor heat transfer	One or more valves closed in primary circuit / physical blockage in primary circuit / insufficient heat transfer fluid / frost damage to collector / incorrect controller settings / incorrect pump speed / incorrect location of sensors / insulation missing
Discharge from safety relief valve	Insufficient expansion capacity in primary circuit / overfilled primary circuit / faulty safety relief valve
Expansion vessel split	Expansion vessel maximum operating pressure has been exceeded
Pipe or fitting failure causing fluid leak	Pressure or temperature exceeds fitting specification / frost damage / rodent attack
Collector temperature over 100 °C	If store temperature has been satisfied, this is not a fault but a DTC warning indicator

Open checklist as PDF document   PDF Link

Click the **'Close'** button to move on   Close

Fault-finding and Rectification Checklist for Fully Filled Pressurized Systems	
Items which differ from the Fault-finding list for Drainback Systems are shown in a different colour	
Symptom	**Fault**
No power at controller	Not enough sun (PV driven)/incorrect wiring/loose connection/open circuit in cable/broken fuse or circuit breaker/open resettable aquastat/controller malfunction
Pump not working	No power from source/wiring incorrect/incorrect control settings/pump seized/pump valves closed/controller malfunction
No flow, poor flow or poor heat transfer	One or more valves closed in primary circuit/air lock/check valve stuck/physical blockage in primary circuit/insufficient heat transfer fluid/open mechanical isolator on AAV/frost damage to collector/incorrect controller settings/incorrect pump speed/incorrect location of sensors/insulation missing
System pressure too low	Fluid leak/safety relief valve has discharged/expansion vessel gas pre-charge has been lost
Discharge from safety relief valve	Insufficient expansion vessel capacity/expansion vessel membrane has failed/blocked pipe leading to expansion vessel/faulty safety relief valve
Expansion vessel split	Expansion vessel maximum operating pressure has been exceeded
Pipe or fitting failure causing fluid leak	Pressure or temperature exceeds fitting specification/frost damage/rodent attack
Collector temperature over 100°C	If store temperature has been satisfied, this is not a fault but a DTC warning indicator

Fault-finding and Rectification Checklist for Drainback Systems	
Items which differ from the Fault-finding list for Fully Filled Pressurized Systems are shown in a different colour	
Symptom	**Fault**
No power at controller	Not enough sun (PV driven)/incorrect wiring/loose connection/open circuit in cable/broken fuse or circuit breaker/open resettable aquastat/controller malfunction
Pump not working	No power from source/wiring incorrect/incorrect control settings/pump seized/pump valves closed/controller malfunction
No flow, poor flow or poor heat transfer	One or more valves closed in primary circuit/physical blockage in primary circuit/insufficient heat transfer fluid/frost damage to collector/incorrect controller settings/incorrect pump speed/incorrect location of sensors/insulation missing
Discharge from safety relief valve	Insufficient expansion capacity in primary circuit/overfilled primary circuit/faulty safety relief valve
Pipe or fitting failure causing fluid leak	Pressure or temperature exceeds fitting specification/frost damage/rodent attack
Collector temperature over 100°C	If store temperature has been satisfied, this is not a fault but a DTC warning indicator

ACTIVITY 54

What actions and choices can you make at the design and installation stage to minimize call backs and faults in the solar system?

CHECK YOUR KNOWLEDGE

1. **Choose the appropriate checklists and the order you would use them during the handover period of the solar domestic hot water system to the customer.**

Options	Order
Fault-finding checklist	
Handover checklist	
Documentation checklist	
Maintenance checklist	
Commissioning checklist	
Installation checklist	
Filling checklist	

2. **Why are installation checks made?**

☐ a. To check the system complies with the design specification

☐ b. To identify any obvious defects before filling

☐ c. To allow for minor changes requested by the customer

☐ d. To check that insulation is installed on every pipe and vessel

☐ e. To check electrical supply from the mains

3. What drainback system measurements should be recorded and handed over to the customer for future maintenance visits?

☐ a. Heat transfer fluid type, frost protection level and total fluid content

☐ b. Solar collector net surface area

☐ c. Expansion vessel gas pre-charge pressure

☐ d. System flow rate and pump speed

☐ e. Safety relief valve discharge rating

☐ f. System operating pressure and expansion vessel gas pre-charge pressure

4. At what times should a hot water cylinder be at its hottest and coolest during the day, and when should it ideally be maintained at 60°C for an hour to minimize the risk from Legionella?

	Hottest	Coolest	At least 60°C for an hour to kill Legionella
Early morning			
Late morning			
Early afternoon			
Late afternoon			
Early evening			
Late evening			

Chapter 9

END TEST

This end test will check your knowledge on the information held within this workbook.

The Test

CHAPTER 1

1. **Which of the following statements describe advantages of solar domestic hot water systems?**

   ☐ a. Reduces heating bills

   ☐ b. Cheap to set up and install

   ☐ c. Reduces carbon emissions

   ☐ d. Cheap to run

   ☐ e. Uses almost no fossil fuels

   ☐ f. Provides 100 per cent of hot water demand

2. **What is considered the best inclined angle for a solar collector in the south of England?**

   ☐ a. 0°

   ☐ b. 15°

   ☐ c. 35°

   ☐ d. 70°

   ☐ e. 90°

3. **Which of the following forms of renewable energy are directly or indirectly powered by the Sun (and sometimes energy from the Moon as well), or do not rely on energy from the Sun at all?**

   Wind power

   Solar power for heating water

   Photovoltaic cells

   Tidal power

   Hydroelectric power

   Geothermal power

   Biofuels

   Wood

   Complete the lists by entering them into the table in the correct columns:

Renewable energy directly powered by the Sun	Renewable energy indirectly powered by the Sun (and sometimes Moon)	Renewable energy not powered by the Sun

CHAPTER 2

4. **How many people must work onsite before a trained first aider becomes a legal requirement?**

   ☐ a. 1

   ☐ b. 50

   ☐ c. 100

CHAPTER 3

5. **Which requirement(s) of BS EN 12975 have to be passed or included before a collector qualifies for BS EN 12975?**

   ☐ a. Durability

   ☐ b. Performance

c. Label on the collector

d. Installer instruction manual

6. **Here are three terms:**

Reflectance

Absorbtance

Radiance

Label each box on the diagram to show which factor is operating on the radiation indicated.

7. **What is the absorber within a solar hot water collector typically made from?**

a. Plastic

b.. Metal

c. Rubber

d. Glass

CHAPTER 4

8. **Solar domestic hot water heating systems must have two storage vessels to keep the solar heated water separate from the water heated by the back up boiler or immersion heater.**

a. True

b. False

9. **Where would you find the following items in the storage cylinder in a solar domestic hot water system?**

The cold water inlet

The hot water outlet

The solar heat transfer coil

The back up heat transfer coil

Tick the appropriate box in the following table:

Item	Top of storage cylinder	Upper middle of storage cylinder	Lower middle of storage cylinder	Bottom of storage cylinder
Cold water inlet				
Hot water outlet				
Solar heat transfer coil				
Back up boiler heat transfer coil				

10. Which of the following statements apply to twin-coil storage vessels? Enter true or false in the correct column in the table:

Statement	True	False
Approximately 1m² of flat plate collector requires a volume of 50 litres of water to heat		
Approximately 1m² of tube collector requires less than 50 litres of water to heat		
Volume of water to be heated by the back up system (Vb) must meet the needs of the household when the solar contribution is low		
The volume of water heated by the collectors added to the volume of water heated by the back up system gives the size of the storage vessel required		

CHAPTER 5

11. Solar primary circuits can be direct or indirect. Identify the characteristics of each by completing the table.

Characteristic	Direct primary circuit	Indirect primary circuit
Solar primary circuits connect the energy generation (the collector) with the energy storage (the hot water cylinder) and provide a means of transferring heat from one to the other		
Heat transfer fluid can be water stored in the hot water cylinder		
Heat transfer fluid flows through a heat exchange coil		

12. Which type of primary circuit are you most likely to encounter in the UK?

☐ a. Indirect fully filled pressurized

☐ b. Indirect drainback

☐ c. Direct

☐ d. Split collector

13. **What are the hazards that need to be addressed by an indirect solar circuit?**

☐ a. Withstanding frost

☐ b. Stopping excessive sunshine

☐ c. Managing high temperatures and steam

☐ d. Limiting the effects of limescale

☐ e. Limiting tree growth and excessive shading

☐ f. Controlling bacterial growth (including Legionella)

☐ g. Managing fluid expansion without user intervention

☐ h. Controlling flow of transfer fluid during hot water use

☐ i. Limiting energy loss

CHAPTER 6

14. **Identify the two main purposes of a controller within a solar domestic hot water system**

☐ a. To maintain user safety and minimize scalding

☐ b. To allow the system to be turned off when the user is away

☐ c. To efficiently control the system circulating pump

☐ d. To control the amount of transfer fluid being circulated

☐ e. To allow the user to adjust the hot water temperature at the taps

☐ f. To set the times when the system is on and off during the day

15. **How does a Differential Temperature Controller work?**

☐ a. It monitors temperatures between flow and return in the solar primary circuit

☐ b. It controls the pump using temperature at the top and bottom of the storage cylinder

☐ c. It controls the pump on/off switch using timers

☐ d. It controls the pump on/off switch using temperatures from the collector and cylinder solar volume

16. **Sensors must be connected to the Differential Temperature Controller using the correct type of cabling. Which cabling would you select for each of the following situations described in the table?**

You have four types of cable to choose from:

☐ a. White heat resistant cable diameter 0.75mm

☐ b. White heat resistant cable diameter 1.5mm

☐ c. Black UV resistant cable diameter 0.75mm

☐ d. Black UV resistant cable diameter 1.5mm

Situation	Type of cable
Collector to DTC, distance – 20m outside, 20m inside	
Top of storage tank to DTC – 45m inside	
Bottom of storage tank to DTC – 55m inside	

CHAPTER 7

17. **What temperature must components of a solar domestic hot water system be able to withstand without distortion or failure?**

☐ a. The maximum temperature experienced in that section of the solar circuit, which is often said to be 150°C for much of the solar circuit

☐ b. 100°C, i.e. the boiling point of water

☐ c. 210°C, the highest temperature in a flat plate collector

☐ d. 300°C, the highest temperature reached in a direct flow evacuated tube

18. **Are drainback vessels to be installed in flow or return pipes?**

☐ a. Always installed in flow pipes

☐ b. Always installed in return pipes

☐ c. Depends on the manufacturer's instructions

☐ d. In neither flow or return pipes but at the end of a long leg

CHAPTER 8

19. **What are some of the hazards that could be encountered during commissioning?**

☐ a. Scalding

☐ b. Chemicals

☐ c. Limescale

☐ d. Legionella

☐ e. Overcast day

20. **What items need to be checked during an initial filling of a new solar domestic hot water system?**

☐ a. Check for leaks

☐ b. Check pump power

☐ c. Check controller power

☐ d. Check system flow

☐ e. Flush out flux and debris

☐ f. Check insulation is thick enough

☐ g. Collector performance

Answers

End Test Answers

1. A, C, D, E

2. C – The best inclined angle for a solar collector in the south of England is 35°

3.

Renewable energy directly powered by the Sun	Renewable energy indirectly powered by the Sun (and sometimes Moon)	Renewable energy not powered by the Sun
Solar power for heating water	Wind power	Geothermal power
Photovoltaic cells	Tidal power	
	Hydroelectric power	
	Wood	
	Biofuels	

4. B – For larger sites of over fifty people, there must also be at least one person with first aid training

5. A, C, D

6.

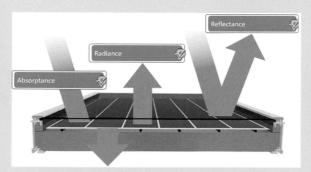

7. B – The absorber within a solar hot water collector is typically made from metal

8. B – False

9.

Item	Top of storage cylinder	Upper middle of storage cylinder	Lower middle of storage cylinder	Bottom of storage cylinder
Cold water inlet				✓
Hot water outlet	✓			
Solar heat transfer coil				✓
Back up boiler heat transfer coil		✓		

10.

Statement	True	False
Approximately 1m² of flat plate collector requires a volume of 50 litres of water to heat	True	
Approximately 1m² of tube collector requires less than 50 litres of water to heat		False
Volume of water to be heated by the back up system (Vb) must meet the needs of the household when the solar contribution is low	True	
The volume of water heated by the collectors added to the volume of water heated by the back up system gives the size of the storage vessel required	True	

Please note that the above answers are generalizations that apply across the whole range of solar thermal systems. Because they are general rules of thumb, always use manufacturer's specific information when designing and specifying the solar system

11.

Characteristic	Direct primary circuit	Indirect primary circuit
Solar primary circuits connect the energy generation (the collector) with the energy storage (the hot water cylinder) and provide a means of transferring heat from one to the other	✓	✓
Heat transfer fluid can be water stored in the hot water cylinder	✓	
Heat transfer fluid flows through a heat exchange coil		✓

12. **A – You are most likely to encounter an indirect fully filled pressurized primary circuit in the UK**

13. **A, C, D, F, G, I – The hazards that need to be addressed by an indirect solar circuit**

14. **A, C**

15. **D – A Differential Temperature Controller uses temperatures from the collector and cylinder solar volume to control the pump on/off switch**

16.

Situation	Type of cable
Collector to DTC, distance – 20m outside, 20m inside	(C) Black UV resistant cable diameter 0.75mm
Top of storage tank to DTC – 45m inside	(A) White heat resistant cable diameter 0.75mm
Bottom of storage tank to DTC – 55m inside	(B) White heat resistant cable diameter 1.5mm

17. A – The components of a solar domestic hot water system should be able to withstand up to 150°C

18. C – The manufacturer's instructions determine whether the drainback vessel is to be installed in a flow or return pipe

19. A, B – Scalding and Chemicals

20. A, B, C, D, E

Activity Answers

Activity 1
Renewables

1. Solar energy
2. Biomass
3. Wind
4. Wave
5. Tidal
6. Geothermal
7. Hydro

Fossil

1. Gas
2. Oil
3. Coal

Other

1. Nuclear

Gas, oil and coal are mainly made up of carbon and so burning these fuels means we create more carbon dioxide. Moving to the hydrogen economy is not easy as hydrogen is a very light and volatile gas and so is not easy to store in fuel tanks. On top of this, we need a lot of renewable or other source of energy to create the hydrogen fuel.

Gas is more environmentally friendly than oil or coal as gas is a mixture of carbon and hydrogen (CH_4) whilst oil and coal contain much more carbon than hydrogen. Nuclear is currently fission-based technology. There is much research into fusion-based power generation. Fusion is the reaction that occurs in the Sun. Most commentators believe that fusion-based power generation is at least 50 years from being market ready.

Activity 2
This is almost a trick question; the word fluid is used in its scientific context to represent a liquid or a gas. The answer is air and water. Water, the liquid form, is normally treated with antifreeze to prevent it turning to ice in the solar system.

And by the way, if you are now complaining that you didn't know that in scientific terms a fluid is a liquid or a gas, please go back to school, don't pass go and don't collect £200. This should be basic knowledge for anyone covering this course.

Activity 3
Seasonal Affective Disorder or SAD is associated with lack of sunshine in winter, and two solutions are using a daylight lamp or getting out for a walk at lunchtime. Vitamin D comes from the Sun. Please note that getting out for a walk might not address serious cases of SAD (which can be a debilitating disease) although it is a good cure for a case of the winter blues.

Activity 4
First of all, it is worth noting that there is no such thing as a typical house and the variety of different property types is significant. Having said this, there are many houses with a pitched roof and probably 100 per cent of these lie in the east to

west orientation and 50 per cent in the south-east to south-west orientation. A pitched roof will often have an angle of inclination somewhere close to the 35° angle of orientation. Therefore, many houses with pitched roofs will be suitable for a solar energy system.

Activity 5

In the diagram shown previously, a boiler is used. This boiler could be gas, oil, log, coal or wood fired. An electrical immersion heater could also be used and today, heat pumps, either an ASHP or GSHP might be found. The very latest heating product on the market is called a mCHP and this is an engine that makes both heat and electricity. All these systems could be used to support the solar thermal system.

Because solar power is strong in summer, a well sized solar thermal system should only need minimal back up or support during the sunniest summer months of June and July, whilst in the middle of winter when there is only about a fifth of the solar energy available in summer, the solar system will need the most support to meet the hot water needs of the household and this support may still be required in spring and autumn, although to a lesser extent than in the depths of winter.

Activity 6

Picking up on the points one-by-one:

Freezing. Frozen pipes are a real problem as frozen pipes split, causing significant leaks. For externally located pipes, they are frequently insulated (or drained) and internally located pipes tend to be located in the heated areas of the house. The house is kept above freezing during the cold season using the thermostats and boiler on frost setting when the property is unoccupied.

Stagnation. This is when the solar system boils the internal fluid. Boiler circuits have safety features on both the boiler and cylinders to prevent this happening. When thinking about the boiler, it features high level trips to prevent the primary water reaching 100°C and an unvented cylinder has three levels of control, a thermostat, a mechanical reset thermostat and a TPRV to prevent the water ever reaching boiling point.

Limescale. We either try to eliminate or manage limescale in all heating systems. Elimination, whether successful or otherwise, is achieved via softeners and conditioners, and limescale is managed by preventing the water going over 60°C in hard water areas. However, most people tend to live with the consequences of hard water rather than eliminating it completely by water softening.

Bacteria. Bacteria can be killed by UV light and other methods. However, in the vast majority of heating systems, most installers set a regular pasteurization cycle to heat the hot water cylinder to 60°C to significantly reduce bacteria concentrations. Heating installers also take other active measures such as eliminating dead legs and changing old open-to-air cold water cisterns for modern insulated and covered cold water cisterns to minimize all bacteria risks.

Scalding occurs when water above 40°C 'burns' or scalds skin. In a healthy adult this can cause considerable distress and in extreme cases, death. In children and old people the consequences are far more serious. Management of the risks comes in many forms, such as warning signs in many commercial premises, to fitting blending valves to 43°C or 48°C as appropriate on different points of use. However, increasing the scalding prevention also reduces the bacteria protection levels and also increases complaints about the water being lukewarm, so all these choices need to be informed to achieve good outcomes. Current reports are saying more people die from scalding as compared to bacterial infections, so this is the higher level of care requirement.

Activity 7

The following list of materials would cover the task:

- A hosepipe, preferably black or as dark as possible to act as the solar collector or absorber plate
- Some insulation to limit the heat escaping from the back of the collector
- Some wood to build a supporting frame and some varnish or woodstain to protect the wood from the elements
- Some glass, preferably toughened to act as the front glazing

You might have chosen to build a metal frame rather than a wooden frame and this, as long as it is corrosion protected, is also a good solution.

Regards passing EN 12975, this would depend on how well constructed the final product is. Toughened glass should help in passing impact tests and good varnish should increase lifetime. The insulation would have to resist fairly high temperatures although a hosepipe, even behind a glass cover, is unlikely to reach above 150°C. The biggest issue with this solution would be the melting of the hosepipe as the internal water reached over 100°C in stagnation mode. However, the hosepipe could be replaced with copper pipe brazed to a copper sheet painted black.

The EN 12975 performance test doesn't have a pass or fail criteria, only a performance test result. However, it is worth noting that a home-made solar collector, unless much care and product development is taken with its construction, is likely to have a much lower performance as compared to a professionally constructed factory made product. Factory made solar collectors normally have low iron glass, selective coating absorber surfaces and high performance insulation and boxes, all of which increase solar performance.

Activity 8

The round shape means that the tubes are very good at managing the huge internal to external pressure difference. There is a whole atmosphere of pressure acting on the outside of the tube and this pressure would probably collapse a square tube.

The main advantage of direct flow tubes is that they can be mounted in any orientation and at any angle. Evacuated tubes have to be mounted vertically at an angle between 15° and 90°. However, if a direct flow tube loses its vacuum or breaks, the whole solar system needs to be drained before the system can be refilled, whereas with a heat pipe and a correctly designed manifold, individual heat pipes can be changed without draining down the solar primary circuit.

Activity 9

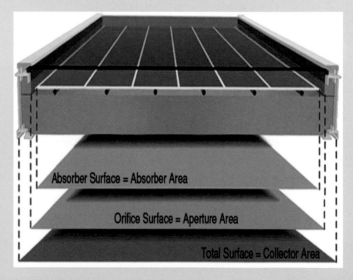

Absorber Surface = Absorber Area

Orifice Surface = Aperture Area

Total Surface = Collector Area

Activity 10

The glazing will transmit light and there will be a transmissivity figure for how much light the glazing transmits through to the absorber area of the solar collector. The higher the level of this transmission, the more solar light and so energy will reach the collector surface. The more energy that reaches the collector surface, the more efficient the collector.

Activity 11

1. At delta 20 K, the unglazed non-selective flat plate is 95%, non-selective surface flat plate is 82%, selective surface flat plate is 81% and selective surface evacuated tube is 78%.

    At delta 54 K, the unglazed non-selective flat plate is 14%, non-selective surface flat plate is 63%, selective surface flat plate is 71% and selective surface evacuated tube is 72%.

Activity 12

1. Heating just a swimming pool for summer use only would probably be most cost effectively achieved with an unglazed solar collector. However, you might choose a glazed collector option if space was at a premium or if appearance considerations came into play.

2. Heating hot water and a swimming pool would normally require a glazed collector, as an unglazed collector would not typically reach the temperatures required to supply hot water heating. Flat plates would probably be the most cost effective option here.

3. Heating hot water and providing some of the space heating load in spring and autumn would normally require a high performance collector such as evacuated tubes, or a high performance flat plate such as a double glazed model. This is because space heating normally requires larger ΔT between the outside air temperature and the flow temperature of the heating distribution system. This of course depends on the particular design of the heating distribution system. However, in most cases a high performance collector would probably be the better option.

Activity 13

1. Single lapped flat or profiled concrete tiles. These tiles are the easiest to remove and work with, so it is easy to either fit a collector above these tiles (on-roof) or to mount the collector in-roof with an appropriate flashing kit. It will probably, depending on how long the on-roof frame takes to build, take a bit longer to mount in-roof. However an in-roof collector arguably looks more attractive and so might be the best option for increasing sales opportunities.

2. Double lapped plain tiles. These tiles are fairly easy to remove. However, they are small and so more fiddly to work as compared to single lapped tiles and when old, they crack easily. Also, in-roof flashing kits are not so commonly designed to work with this type of tile (these flashing kits sometimes have a secret gutter to improve performance). Therefore, mounting above

roof is more common when working with this variety of tile.

3. Double lapped slate roofs. These tiles are difficult to remove as a slate ripper is required to remove the old heads of the fixing nails and a large section of roofing is both difficult to uncover and then easily refit the slates without leaving lots of tails. Therefore, it is infrequent to mount collectors in-roof on slates unless the project in question is a new build or re-slating project.

Activity 14
In a domestic property, the hot water might be required at any point in the day or night, whether the Sun is shining or not. However, a solar water heating system will only be capable of heating the hot water at a steady and fairly slow rate throughout a sunny day. Therefore, without a hot water store that can set aside the solar energy for use when the household occupants need the hot water, the domestic solar system would be of no use to the occupant. The back up heating source is required because it is impossible to guarantee economically storing enough solar hot water in a northern European climate and so the hot water system will need an additional source of heat to support or back up the solar heating system.

Activity 15
Because the solar gain occurs during the day, leaving a cylinder full of cool water for the Sun to heat during the day will maximize the solar gain and then using the water during the evening or first thing in the morning (a modern well insulated solar system should retain most of its heat overnight) will ensure that there is a significant volume of cool water for the Sun to heat during the day.

Likewise, timing of the back up heating system is also important. The back up system should ideally be turned on at the end of the solar day so that the back up volume is brought up to pasteurization temperature once a day and there is a good volume of hot water at the top of the cylinder to keep the household occupants supplied with enough hot water for their needs. Turning the back up heating system on in the morning after everyone has been to the bathroom means that the top of the cylinder is full of hot water, so the Sun can't heat this volume.

Activity 16
An unvented cylinder stores a large body of water at close to 100°C. If the cylinder passes boiling point and the cylinder breaks, this whole body of water will flash to steam, expanding 1600 times in volume in the process and creating a large explosion which would normally bring down the property and perhaps neighbouring buildings. If you search the Internet for examples of unvented cylinder explosions, you will see some dramatic examples of this happening in controlled circumstances.

Therefore, safety systems are put in place to manage this process. Across the UK and Ireland, the three levels of safety set out above are required. The advantage of this high level of safety is just what it says on the tin, it is a high level of safety that makes an explosion even less likely. However, it costs more for society to implement, install and maintain these higher levels of safety and so getting the optimum level of safety is the key solution in any situation. However, as the optimum level of safety is always an opinion-based process, safety levels will always cause considerable debate and discussion.

Activity 17
When a collector comes out of stagnation, there is still pressurized steam in the solar circuit and this steam can travel down the flow line all the

way to the cylinder heat exchanger where it will probably condense. A central heating two-way valve is not designed to handle pressurized steam and so this could fail. If this valve fails, there is no longer any second level of control on the cylinder. Therefore, it is outside of spec to fit central heating valves in solar circuits and would contravene building regulations.

Activity 18

If the primary water in the thermal store is kept at 80°C and the secondary water is delivered to the taps at 60°C, then the thermal heat exchanger can be much smaller than if the primary water is stored at 70°C in the thermal store. This is because of the greater temperature difference.

A thermal store is a form of instantaneous water heater in that it heats much of the water delivered to the taps from cold to hot within the heat exchanger volume. Therefore, the secondary water must have enough residency time in this heat exchanger volume to reach delivered-to-taps temperature. When designing vented or unvented cylinder heat exchangers, the designer must make sure that at the central heating primary circuit flow rates and temperature differences, the heat exchanger coil is big enough to heat sink all the heat.

A bigger coil with longer residency time and greater surface area is normally required in a thermal store to make sure the inlet water rises from 10°C to 60°C. In a vented or unvented cylinder, the temperature difference will typically be a flow temperature of 70°C to 80°C and a return temperature of 50°C to 70°C. The significantly smaller temperature differences with a vented or unvented cylinder are why a smaller heat exchanger can normally be used in these cylinders.

Activity 19

Legionella is pasteurized after a few minutes at 70°C. However, the secondary tap water is not likely to be resident within the combi boiler for enough time to become fully pasteurized, especially as it is rare for water leaving a combi boiler to reach 70°C. Therefore, extra precautionary measures to fairly regularly pasteurize or otherwise manage bacteria issues in the solar preheat store should normally be considered for this type of solar design.

Activity 20

Twin-coil

V_b will equal 130 litres. The house will have occupancy of 5 (number of bedrooms plus 1) so V_d will equal 5 * 45 = 225 litres. V_s is:

- either 80 per cent V_d = 180 litres
- or 5 (occupancy) * 25 litres/person = 125 litres

Therefore V_t minimum = 255 litres of which 130 litres is for V_b and 125 litres for V_s

Two cylinder

V_b will equal 130 litres. The house will have occupancy of 5 (number of bedrooms plus 1) so V_d will equal 5 * 45 = 225 litres. V_s is:

- either 80 per cent V_d = 180 litres
- or 5 (occupancy) * 50 litres/person = 250 litres

Therefore V_t minimum = 310 litres of which 130 litres is for V_b and 180 litres for V_s

Activity 21

Flow rate/m^2 = 2 litres/min divided by 4m^2 = 0.5 litres/min

Minimum coil surface area = 4m^2 collector * 0.1m^2 coil/m^2 collector = 0.4m^2 coil area

Activity 22

If 90 per cent of the bacteria is killed after 32 minutes, twice as long, that is 64 minutes, will kill 90 per cent * 90 per cent of the bacteria which will be 99 per cent of the bacteria. Therefore, 1 per cent of the original bacteria that was in the original sample will remain. In most circumstances, leaving just 1 per cent of the original bacteria would not do anyone, including old people and young children any harm. However, if the original sample was very dirty, then this 1 per cent would be enough to harm someone, especially anyone susceptible to Legionella bacteria. Therefore these bacteria risk assessments must be thorough and carefully conducted. A well designed hot water system will keep the occupants with a very low risk of Legionella and other bacteria.

Activity 23

The cylinder stat is normally a bimetallic strip that switches the boiler off at the present cylinder temperature. The sensor for Tmax on the solar controller is normally a PT1000 which is a resistance sensor. The solar controller measures the resistance across the PT 1000 and so knows the current operating temperature at this sensor location within the cylinder.

Activity 24

They only use potable grade materials in the solar circuit so that no contamination is brought into contact with the hot water that goes to the taps. This often means that they usually use a heat exchanger in the collector to maintain the water quality.

Activity 25

The answer to these various points is provided in the subsequent section of the workbook

Activity 26

You need to ask the solar collector supplier for the internal volume of the solar collectors used in the design and then add to this the relevant length of the pipework, multiplied by the internal area of the pipes. Don't forget to add a few extras litres to the final result as a safety factor.

Activity 27

This is similar to the calculation of the expansion vessel volume for a fully-filled circuit. You need to ask the collector supplier for the internal volume of the collectors, add to this the internal volume of all the pipework above the collector and finally allow for the air volume at the top of the drainback cylinder. The drainback cylinder supplier should provide you with this air volume. Finally add a few extra litres as a safety factor.

Activity 28

A drainback solar circuit has fewer components and a simpler layout. It also manages stagnation and freezing with more elegance as compared to a fully-filled circuit. However, there is typically a greater volume of fluid in a drainback solar circuit and this will make this system a bit less responsive. A drainback solar circuit needs to have a fall on all the pipework above the drainback vessel and so this makes both collector and pipework design less flexible. In truth, if one layout was much better than the other, then this layout would dominate and be the only one on the market.

Activity 29

This example has been chosen to offer very little between the two quotes. Costwise the difference is negligible so unless the householder is very cost driven, this can be discounted as a contributory factor in the choice of system.

The east–west system has the advantage of being able to collect some heat throughout the day, as a pure east-facing system will lose the late afternoon solar thermal energy. However, if there is good morning sunshine, then the east-facing system will collect more heat in the morning as it has a bigger surface area.

Looking at surface area, 5m^2 of flat plates will probably collect a bit more energy over the year as compared to 4m^2 of tubes, but this extra advantage will be negligible and the tubes will gain a bit more energy in the depths of winter.

So the final choice, if the quotes feature high quality flat plates compared to high quality evacuated tubes, is really down to appearance and whether the customer wants to have the flat plate or tube 'look' on their roof.

Activity 30

The flow rate should be set to 3.6 litres/min (4 * 0.9).

This system should use setting 1 slow on the pump and consume 40 watts of power.

A modern high efficiency solar pump would probably use considerably less energy as compared to the standard solar pump in this example. Many pump manufacturers have developed special solar pumps over the last few years that use considerably less energy than standard solar pumps.

Activity 31

Fundamentally, the answer to both questions is yes. The DTC we looked at before we discussed thermosiphon solar systems was an electronic controller. A thermosiphon system uses a mechanical-based gravity control mechanism and this system will be a form of differential temperature control as the water in the collector needs to be warmer than the water in the storage cylinder for circulation to occur. Indeed, this is an elegant form of control in that the greater the temperature difference, the greater the flow rate. In traditional engineering, nearly all control systems were mechanical. However, today electronics has taken over most control functions from mechanical systems. It's just that in sunny, warm climates, it is more cost effective to use solar thermosiphon systems and in temperate climates, electronically controlled and electrically pump driven solar systems, and this is why we use either system.

And a thermosiphon system will normally be interlocked in that in most designs, it would be impossible to export back up heating to the ambient environment through the solar collector. The only rider to add here is to emphasize the word 'normally' because we expect that if someone put their mind to it, they could design or specify a solar system which could lose back up heating to the local environment. For example, in certain cases, without check valves, the thermosiphon system could run backwards during the night and so be exporting back up heating to the ambient environment.

Activity 32

Two separate control circuits means added complexity and cost, so if a programmer was mass produced to cover both central heating and solar heating in one system, it should prove to be lower cost to produce and install. These cost benefits would probably be only properly realized if the whole heating system was fitted at one time. There would also be benefits in operation of the hot water system as it would be easier to programme and automate the back up on and off times to maximize solar gain if the two control circuits were operated from one central programmer. Several heating control suppliers now provide controls that integrate the two control systems in one programmer and it is anticipated that this will be an ongoing growth trend.

Activity 33

1. 1097 ohms = 25°C

2. 1366 oms = 95°C

3. Infinity ohms indicates that the sensor has failed and probably needs to be replaced

Activity 34

Unfortunately, this discovery is more common than we would like it to be. Hopefully the tail of the external solar sensor is available in the loft or other accessible place so that you can connect a suitable cross sectional area cable from the loft through to the DTC. We also hope that the telephone cable is in a readily accessible place so that it can be quickly and easily replaced. Make sure you properly fix the new cable in place. The worst case scenario is if the external tail of the solar sensor is not available, in which case you will need to obtain access to the roof with all the health and safety implications this entails to attach the new cable to the tail of the solar sensor.

Telephone cable, because it is so thin, will probably increase the resistance of the solar sensor circuit and so cause this sensor to give a greater temperature reading than the actual temperature of the collector.

Activity 35

1. The solar pump will be running and the system collecting heat as $T_{collector} - T_{cylinder} > 7°C$ temperature difference ($63 - 55 = 8 > 7$ (ΔT_{on} 7°C))

2. The solar pump will be off and the system will have stopped collecting heat as $T_{collector} - T_{cylinder} < 4°C$ temperature difference ($62.9 - 59 = 3.9 < 4$ (ΔT_{off} 4°C))

3. The solar pump will start running again and the system collecting heat as $T_{collector} - T_{cylinder} > 7°C$ temperature difference ($66.1 - 59 = 7.1 > 7$ (ΔT_{on} 7°C))

4. Even though the temperature difference is greater than 7, the solar pump will stop as Tmax has been reached and the system will not be collecting heat. Tmax = 60°C even though $T_{collector} - T_{cylinder} > 7°C$ temperature difference ($67.1 - 60 = 7.1 > 7$ (ΔT_{on} 7°C))

5. The solar system will switch back on and the solar pump will be running as Tmax < 60°C and $\Delta T_{on} > 7°C$ ($85 - 55 = 30 > 7$ (ΔT_{on} 7°C)). The solar system will probably be running for a while as the solar collector is fairly hot and the solar circuit will take a while to cool the collector down.

Activity 36

Example Answer

There are many ways this answer could be covered. The author would choose to start with:

- Primary
- Secondary

And then look at sub-functions within these two categories. For example:

Primary. This circuit collects the solar energy. It consists of four main elements:

- The collector
- Flow line
- HW store (cylinder) heat exchanger
- Return line

Flow line of the primary circuit

The collector (and parts of the flow and return line) might also be exposed to the external environment so these sections will need to be wind, rain and weatherproof (which includes UV protection) and also maximize solar gain at economic cost.

Thinking directly about the flow line, this element has two extreme conditions, exposure to frost and exposure to the maximum temperature in the collector. Frost is only likely to occur in the external elements and close to external elements in the flow line. It can be managed in one of three ways:

- Make sure that the liquid in the collector doesn't freeze (use appropriate concentration of antifreeze)
- Allow the liquid in the collector to freeze and then make sure that the components in the flow line can resist this freezing
- Replace the liquid in the collector with a gas that doesn't freeze (the drainback option)

The maximum temperature in the collector depends on the collector type. For example, a good flat plate will create up to 210°C whilst a direct evacuated tube will create up to 300°C. When the solar system comes out of stagnation mode, the flow line of the solar system will be exposed to these temperatures for a short time. Therefore, these components must be able to withstand these short blasts of high temperature without significant degradation. All fittings should be of brazed or compression or similar design that can withstand short blasts of high temperature, and the pipework, insulation and any fittings must also be spec-ed to deal with these high temperature situations.

All the above is in addition to the standard function of the flow line, which is to efficiently and effectively transport the heat from the collector to the solar store with minimal heat and pressure loss to maximize solar gain and minimize pumping loss.

Activity 37

As solar collectors and systems have improved, due to the higher temperatures generated in these circuits, a higher specification of jointing was required to join different sections of pipe and components together to make up the solar circuit. Lead-free soldered joints can still be used if recommended by the manufacturer because these solar collectors and systems will not exceed the specification of these fittings

Activity 38

The secondary hot water circuit doesn't reach the high temperatures experienced by the solar primary circuit. Likewise, the back up heating system also doesn't reach these high temperatures so polyethylene insulation can be used anywhere on the heating system except for the solar primary circuit. With of course the additional rider that the manufacturer or supplier might have their own recommended solutions for their heating system and you should always follow their instructions.

Activity 39

Foil-faced mineral or glass wool is typically specified for plant rooms and commercial installations due to its high performance and fire resistance. However, it takes longer to fit as compared to the very flexible HT elastomeric insulation, so this latter insulation is normally used on domestic installations.

Activity 40

1. We are being conservative in the text. Manufacturers will always be improving their product ranges. We will use base rather than

high performance results in our information to make sure we adopt a conservative base case.

2. Water expands 1600 times when it turns to steam. It is impossible to estimate how much water will turn to steam. However, we know that:

    a. In stagnation, steam can be in the collector and the first two vertical metres of pipework

    b. Coming out of stagnation, steam is typically in the collector and the whole of the flow line

If both volumes are calculated and the result multiplied by two and the tidal volume of the expansion vessel is included, the expansion vessel will be just big enough for this circuit. Therefore add a few extra litres by going to the next size vessel and this vessel will adequately cover the needs of this solar circuit.

3. Poor emptying behaviour is created by top rather than bottom connections and long vertical pipe runs within the collector. Therefore, top manifold direct vacuum tubes should ideally be located in a horizontal rather than vertical manner. This does not apply to heat pipe vacuum tubes that will, unless unnecessarily restricted, automatically act as good emptying collectors. The Hausner and Fink paper highlights in more detail good and poor layouts. The expansion vessel should be located above the check (non-return) valves so that the solar collectors can easily empty and fill the expansion vessel volume.

Activity 41
The drainback vessel should have:

1. An air pocket at the top of the drainback vessel

2. Enough tidal volume to accommodate filling all the air spaces above the drainback level

3. Enough water volume at the bottom of the vessel to accommodate for some inevitable fluid loss that occurs in all sealed circuits

The tidal volume is the only element of the system that must be accurately calculated and it consists of all the internal volume of the pipework and collectors above the drainback vessel. The size of the air pocket and bottom water volume are judgement calls. On a domestic system, it is fairly common to find a ten litre drainback vessel that holds one to two litres of air and so has eight or nine litres of liquid volume. Some drainback suppliers state that only water should be used as the transfer fluid. However, if any of this water ends up 'trapped' in the collector during an overnight frost, it can burst an internal collector pipe, so if in any doubt, use antifreeze in the drainback circuit.

Activity 42
Ideally, if the air separator can withstand steam, the air separator can be fitted on the flow line fairly close to the collector. However, it doesn't need to be fitted here and it can be located in any position on the circuit where it assists in removing air bubbles from the antifreeze mix. Please note that antifreeze tends to entrap far more air than plain water and so more care is required in purging solar systems from air. The most important issue is to remember to mechanically isolate the air vent from the circuit after commissioning so that during stagnation, steam can't escape from the solar circuit.

Activity 43
A 4.2m² collector with a recommended flow rate of 1 litre/minute/m² collector would use a flow rate of 4.2 litres/minute. Higher flow rates will

keep the collector cooler and so improve collector efficiency. However, greater flow rates will also increase pumping losses and also cause the system to cycle on and off, so decreasing system efficiency more. Therefore increasing the flow rate is a situation of diminishing returns, and optimum flow rates achieve a balance between these effects.

Activity 44

The connection tails of the solar collector can be extremely hot during sunlight or daylight and so these ends can be wrapped in a cloth during fitting to avoid accidental blister burns if these ends touch skin. Obviously, these cloths must be clean and smooth so that no material is left behind in the primary circuit. If commissioning the solar system during sunlight, the collector can be covered to keep it cool.

Activity 45

The expansion vessel should be mounted on the return line above the return line check valve and the vessel should be mounted vertically downwards with a reasonable run of un-insulated pipework. The pre-charge pressure can be checked with a car tyre pressure gauge before the circuit is filled with fluid. The DTC is not directly mentioned in the above list (although the list points at this item). We would recommend checking the DTC settings against the manufacturer's instructions so that ΔTon, ΔToff and Tmax are all set at this stage and any extra settings (such as programme type) should be selected at this stage of the installation process.

Activity 46

A drainback circuit should only be filled to its fill point in the drainback cylinder. However, a machine filling process could be used as long as after machine filling, there is a big enough reservoir under the drainback cylinder fill point to capture the total air volume of the drainback circuit and the circuit is left to empty to this level after filling from the machine.

Activity 47

Primary circuit should be checked to 1.5 times its maximum operating pressure. Therefore a solar primary circuit with three bar or six bar safety relief valve should be tested to 4.5 bar and nine bar respectively. These test pressures could be achieved if the safety relief valve was blocked off.

Activity 48

COSHH stands for Control of Substances Hazardous to Health. The waste antifreeze must be disposed of in accordance with local authority requirements and would normally be treated as licensed hazardous waste.

Activity 49

The easiest way to avoid entrapped air is to use pre-mix antifreeze that has been stored without being shaken or stirred. If using mixed antifreeze, it is better to mix the solution at the start of the installation and commissioning day and then leave the solution to settle and release as much entrapped air as possible.

Activity 50

A typical 5m head circulating pump uses somewhere between 35 and 45W on setting 1. A low energy solar grade solar pump that is fairly new on the market should use considerably less energy than the typical 5m head circulating pump and pump manufacturers are frequently improving the energy efficiency of their pumps.

It's certainly worth checking that all the collector, cylinder, etc. labelling and all documentation is in full order at this stage so that the handover stage of the installation process runs smoothly and you are professional in your dealings with the customer.

Activity 51

The DTC controller would require a sensor in the back up volume (Vb) and this sensor allied to the programming functionality in the DTC would need to provide time and temperature (thermostatic control) as the two parameters for the back up boiler or other alternative heat source.

Activity 52

The householder normally likes and should be able to view the temperatures in the different parts of the solar circuit, such as top and bottom cylinder temperature and collector temperature. They also like to know the run hours of the solar system. However, it is better to limit access to adjustment settings such as ΔTon, ΔToff and Tmax so that the householder does not accidentally re-programme the DTC settings. Most controllers 'hide' these latter settings so that without prior knowledge, the householder can't alter the programmed settings on the DTC but can still view all the system information.

Activity 53

First of all, ask the householders whether they are pleased with the solar system and whether they think it is making enough energy. Then check the DTC for run hours (in south-east England, a solar system will typically work for 5.5 hours/day over a year, running less hours/day in winter and more hours/day in summer). The run hours will give some indication as to whether the system is working well or not. Check the flowmeter when the system is on to make sure that this is at the right setting and the pressure gauge increases in pressure when the system is in operation. Also confirm with the householder whether the system requires regular top ups or if it runs without any interference. This doesn't take away from any of the list above; it just lets you have a good idea as an installer whether the system is working well or not.

Activity 54

Good design and installation with high quality components improves reliability and so minimizes call backs and faults. Good design includes carefully matching the system size to the property so that the system only stagnates the minimum number of times in summer and also specifying and installing the solar system so that the collector easily empties of liquid during stagnation. This maximizes solar transfer fluid (antifreeze mix) life and so improves reliability of the system. Likewise, specifying good quality components improves life and so reliability. And care taken during installation of the solar collectors means the roof will maintain its integrity for longer. The solar system should have a similar life to the existing roof covering and so sloppy roofing workmanship will mean that the roof is not as weatherproof as it was before the collectors were installed.

Check your knowledge answers

Chapter 1

1.

Photovoltaic (PV)	
Passive solar	
Active solar	

2. A Solar fluid expansion

   C High solar collector temperatures

   D Scalding

   E Stagnation

   F Bacteria growth (Legionella)

   G Limescale

   J Freezing

3.

Hazard	Solution	Answer
Steam & scalding	The components must be able to withstand vapourization and condensation cycles, but also capable of operating without failure or distortion	True
Frost & freezing	Provide low level heating to prevent ice forming in pipes	False
Bacterial growth	Make sure the temperature of the water is raised to 45°C at least once a day	False
Limescale	SDHWS can only be installed in soft water areas as limescale will not form in the pipework	False

4. (a) A sunny day in summer

   (c) A fairly cloudy day in spring

   (d) A bright, cold day in winter

5.

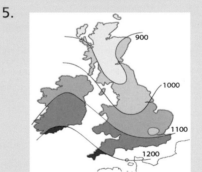

Chapter 2

1. Yes – PPE should still be worn

2. A – You must take responsibility for your safety, however the law requires employers to assess risks, set up control measures and ensure good work practice

3. Plasters, Eye wash, Adhesive dressings, Sterile eye pads, Sterile bandages, Sling, Sterile wound dressings – you should probably add

small items of equipment such as safety pins, disposable gloves, tweezers and scissors. However, please note that tablets or medicines should not be part of a first aid kit, as if used improperly they could cause further harm.

4. D, A, E, C, B

5. C – Powder

6. C – There are three faults: the fuse does not match the rating plate, the cable has been repaired, and there are burn marks on the casing

7. C – 50 per cent

8.

Class of Fire	Fire Extinguisher
Class A – wood paper, textiles, etc	
Class B – oil, petrol, paint, etc	
Class C – gas, acetylene, butane, etc	
Class D metal, magnesium, aluminium, etc	

Chapter 3

1. Heat pipe tube collector

2.

Durability Testing	Performance Testing
Ability to resist leakage	Ability to record efficiency
Ability to resist distortion from internal pressure	Ability to record energy output
Ability to withstand high temperature without fluid	
Ability to resist frost	
Ability to resist rain penetration	

3. B Ambient temperature and collector temperature

4. A heat pipe evacuated tube collector must be mounted with a minimum angle of 25° from the horizontal.

Chapter 4

1.

Unvented	Vented
Can deliver hot water at mains pressure to all outlets	Have a cold water tank in the attic
Have strict regulations controlling their construction and installation	
Do not need supplementary pumps	

2. A All energy sources to have 3 levels of safety

Chapter 5

1. 1. Air separator
   2. Differential Temperature Controller (DTC)
   3. Temperature sensor
   4. Circulating pump
   5. Discharge container
   6. Expansion vessel

2. 1. Differential Temperature Controller (DTC)
   2. Drainback vessel
   3. Safety relief valve
   4. Discharge container

Chapter 6

1.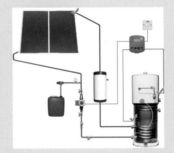

2.

Action	Setting
Activates the pump when the temperature between the collector and the bottom of the storage vessel is more than a set value	T on
Deactivates the pump when the temperature between the collector and the bottom of the storage vessel is less than a set value	T off
Deactivates the pump when the storage tank reaches maximum temperature	T max

Chapter 7

1.

Type of pipe	Fully filled pressurized solar circuits	Drainback solar circuits	Not typically used in solar circuits
Plastic			✓
Copper	✓	✓	
Stainless steel	✓		
Flexible stainless steel	✓		
Barrier			✓

Note: Some manufacturers' use silicone or other pipe materials. If the manufacturer recommends these materials, then always follow their instructions.

2. 1. Safety relief valve
   2. DTC
   3. Pressure gauge
   4. Pump
   5. Thermometers

3.

Component	Standard	Additional	Not Applicable
Air separator		✓	
Check valve(s)	✓		
Circulation pump	✓		
Drain point			✓
DTC		✓	
Expansion vessel connection	✓		
Expansion vessel			✓
Fill point	✓		

Flow regulator	✓		
Heat exchange coil			✓
Pump isolation valves	✓		
Pressure gauge	✓		
Safety relief valve	✓		
Thermometer(s)	✓		

4. B Lead solder fittings

   D Plastic fittings

   E Polyethylene insulation

5. A Positioned at the end of a downward pipe

   B Positioned at the end of a long leg of un-insulated pipework

Chapter 8

1.

Options	Order
Faultfinding checklist	1. Installation checklist
Handover checklist	2. Commissioning checklist
Documentation checklist	3. Handover checklist
Maintenance checklist	
Commissioning checklist	
Installation checklist	
Filling checklist	

2. A To check the system complies with the design specification

   B To identify any obvious defects before filling

3. A Heat transfer fluid type, frost protection
       level and total fluid content

    B Solar collector net surface area

    D System flow rate and pump speed

    E Safety relief valve discharge rating

4.

	Hottest	Coolest	At least 60°C for an hour to kill Legionella
Early morning			
Late morning			
Early afternoon		✓	
Late afternoon			
Early evening	✓		✓
Late evening			

Glossary

Absorber surface area The surface on the collector where solar energy is absorbed.

Absorptance This is the amount of solar radiation being absorbed by a solar collector, typically more than 90 per cent

Active solar heating Active solar heating involves a solar collector mounted in such a way as to gather the maximum amount of solar energy. This is mostly used in domestic hot water systems and for heating swimming pools. Active solar heating means that the system is pumped rather than gravity driven

British Standard The British Standards Institution (BSI) sets quality standards and standard dimensions for equipment and materials. All British Standards start with the letters BS followed by a number.
BS EN 12975:2006 Thermal solar systems and components – solar collectors is in two parts, durability and performance of the solar collector. BS5918, code of practice for solar heating, is also relevant

Cable A conductor used to carry current around an installation. Cables are identified by the colour of the installation

Collector The panel which collects solar energy and can be made from a variety of materials

CSCS Construction Skills Certification Scheme

Dektite Lead flashing containing a silicone rubber bonnet – holes are cut in the bonnet for the pipes to pass through

Differential Temperature Controller This controller operates by monitoring the three key temperatures via the sensors and turning the pump on or off as appropriate. It requires temperatures to be set to activate and deactivate the pump. These key temperatures are 'T on' (also called ΔT on), 'T off' (also called ΔT off) and 'T max'

Diffuse solar radiation Diffuse solar radiation gives us daylight and it can be quite strong. When the Sun's rays hit the Earth's atmosphere some radiation is scattered in all directions creating diffuse radiation

Direct flow evacuated tube collector Type of solar panel. The heat transfer liquid passes through the manifold pipe and each of the evacuated tubes in the assembly

Direct solar radiation Direct solar radiation is sunshine, which is strongest in the summer at midday because then the angle of the Sun is at its highest

Direct systems Have water from the domestic hot water store circulated through the collector then stored until used. The circulating fluid and domestic hot water are the same

Drainback solar primary circuit The main feature of the drainback primary circuit design is that the components are specifically positioned to manage temperature extremes and energy loss

Elastomeric insulation This type of insulation for pipes is not high temperature tolerant or UV resistant, although it is water resistant. HT elastomeric insulation is high temperature tolerant, water and UV resistant and should be used instead

Evacuated tube collector Evacuated tube collectors are made up of a series of evacuated (empty) tubes connected to a manifold pipe assembly, which is all housed in a well insulated metal box. Evacuated tube collectors are nearly always installed on top of roof tiles

Expansion vessels This is a vessel with a membrane inside, which can compensate for volume changes in the primary circuit due to fluid expansion. The vessel is gas filled, and as the fluid expands the gas is compressed, then as the fluid returns to previous volumes the gas pushes the fluid back into the pipe-work. The drainback system does not operate under pressure and is normally a sealed system

Flat plate collector Flat plate collectors are glazed and insulated and these are used for solar hot water heating systems. Unglazed flat plate collectors are often used to heat swimming pools but not for hot water systems

Flow rate The rate at which the heat transfer fluid travels around the circuit. The flow rate needs to be set according to the manufacturer's instructions and the pump needs to be set at the lowest setting that can achieve that rate

Flow regulator Controls the flow in the primary circuit and there is an optimum rate of flow for heat transfer fluid passing through the solar collector

Geothermal Commonly found in volcanic areas such as Iceland where it is used to heat the country's houses

Glazing Glazing materials for the collector need to let the maximum solar energy through to the absorber and the minimum to be transferred back to the atmosphere

Guidance notes There are eight guidance notes to the 17th edition IEE wiring regulations published by the Institution of Electrical Contractors

Hazardous substances A substance hazardous to your health and can be broken down into four main categories –
Irritants – such as soft solder flux – if they come into contact with the skin, eyes, nose or mouth, they can cause inflammation or swelling
Harmful substances – such as lead – can cause death or serious damage when inhaled, swallowed or absorbed by the skin
Corrosive substances – such as sulphuric acid – may destroy parts of your body if they come into direct contact with them
Toxic or very toxic substances – such as bleach – can cause death or serious damage when inhaled, swallowed or absorbed by the skin

Health and Safety at Work Act (1974) (HASAWA) All employers are covered by the HASAWA, which places specific duties on both employers and employees to ensure that workplaces are safe. Non-compliance can result in fines

Heat pipe evacuated tube collector Each pipe is a sealed unit with a large heat transfer 'plug' at the top

Inclination The tilting of something away from a line or surface, or the degree to which it is tilted

Indirect systems This system has water from the domestic hot water store heated by the primary circuit passing fluid through the collector and then transferring heat to the hot water store through a heat exchanger (e.g. coil in storage vessel)

Information signs These are green squares with white symbols to give information

Installation checks Are made to ensure the installed system first complies with the original design specification for the customer, and then checked for any obvious defects

Joints for cables For non-flexible cables, joints may be made by soldering, brazing, welding, mechanical clamps or using compression joints. The jointing device should be selected with reference to the selected cable size

Joule The unit used to measure the amount of work done and used in the formula: work done (J) = force (N) × distance (m). Power is measured in joules per second or J/s and known as watts (W)

Legionella A disease that can form in pipes and that likes temperatures around 20°C to 50°C, particularly around 38°C to 40°C, with other factors such as nutrients and being in slow moving or stagnant fluids. Copper pipes and chorine are toxic to Legionella. This disease can kill

Litmus paper To test for pH of transfer fluid (it should be neutral)

Mandatory signs These are blue circles with white symbols, tell you what you **must** do

Passive Space Heating This uses glass in the building design to heat the rooms. Thermosyphon solar systems are passive

Photovoltaics Directly converts the solar radiation into electricity

Periodic Inspection and Testing It is essential to conduct periodic inspection and testing as the condition of electrical installations deteriorates over time owing to wear and tear, accidental damage and corrosion. The Electricity at Work Regulations 1989 requires that systems are maintained to prevent danger, as far as reasonably practicable

Permit to work A document that specifies the detail of work to be done, when it is going to be done, the hazards involved and the precautions to be taken, and it authorizes the people to be involved in the work

PPE Personal Protective Equipment

Preheat Volume (V_S) The volume heated by solar energy should be at least 80 per cent of the household needs and the pattern of usage affects its efficiency

Prohibition signs These are circular with red crosses through them; they tell you **not** to do something

Radiance This is the amount of absorbed energy radiated back out of the collector, typically between 4 per cent and 12 per cent

Reflectance This is the fraction of light deflected away when it strikes the glazing of a collector

Refractometer Used to determine whether the heat transfer fluid has the correct concentration of antifreeze

Respirator Filter masks stop dust but are useless against gases or vapours, for which you must use a canister respirator

Risk Refers to how likely it is that a potential hazard will actually damage your health

Risk assessment Identifying hazards in the workplace then deciding who might be harmed and how

SDHWS Solar Domestic Hot Water System

Solar Volume (V_S) The volume heated by the solar system must match the surface area of the collector. Approximately one square metre of flat plate collector requires 50 litres of water

Solar primary circuit Solar primary circuits connect the energy generation (the solar collector) with energy storage (hot water cylinder) and provide a means of transferring heat from one to the other

Solar primary circuit controls There are three types of controls – Differential Temperature Controller (DTC) or light intensity sensors, Photovoltaic cells to power the pump, or Thermosiphon primary circuits which use gravity – not typically used in UK

Split collector solar primary circuit Used where collectors need to be mounted on separate sections of roof (east/west-facing roofs)

Stagnation mode When there is no fluid movement, sometimes in summer, the controller switches the solar pump off

Twin-coil storage vessel The twin-coil storage vessel has one coil heated from the solar source, and another heated by the back up heating system, to be used when the solar contribution is low

Thermosiphon effect (also spelled thermosyphon) This is where the hot water is drawn back to the collector and is prevented from doing so by using check valves in the circuit, only allowing flow in one direction

Unvented storage vessels Unvented storage vessels do not have a cold water tank in the attic and can deliver hot water at mains pressure to all outlets. Unvented systems are subject to greater regulations than vented systems. Twin-coil, and single-coil storage vessels can be manufactured for vented or unvented solar hot water systems

Vented storage vessels These have a cold water tank in the attic

Warning signs These warning signs are yellow triangles with black symbols and give notice of a particular hazard or danger

Index